International Conference on Mechanical Engineering, Energy, and Advanced Materials (ICMEEAM 2018)

IOP Conference Series: Materials Science and Engineering
Volume 606

Kota Kinabalu, Malaysia
29 - 30 November 2018

Editors:

Seock-Sam Kim
Willey Liew Yun Hsien
Mohd Azlan Bin Ismail
Md. Mizanur Rahman
Mohd. Suffian Misran

ISBN: 978-1-5108-9483-9
ISSN: 1757-8981

Printed from e-media with permission by:

Curran Associates, Inc.
57 Morehouse Lane
Red Hook, NY 12571

Some format issues inherent in the e-media version may also appear in this print version.

Additional copies of this publication are available from:

Curran Associates, Inc.
57 Morehouse Lane
Red Hook, NY 12571 USA
Phone: 845-758-0400
Fax: 845-758-2633
Email: curran@proceedings.com
Web: www.proceedings.com

Table of contents

Volume 606

International Conference on Mechanical Engineering, Energy and Advanced Materials (ICMEEAM 2018)

29–30 November 2018, Kota Kinabalu, Malaysia

Accepted papers received: 1 August 2019
Published online: 10 September 2019

Preface

Papers

Advanced Materials

Mechanical Engineering

IOP Conf. Series: Materials Science and Engineering **606** (2019) 011001 doi:10.1088/1757-899X/606/1/011001

International Conference on Mechanical Engineering, Energy and Advanced Material (ICMEEAM – 2018)

ORGANIZING COMMITTEE

ADVISOR
Prof. Ir. Dr. Abdul Karim Bin Mirasa
Dean
Faculty of Engineering

CHAIRMAN
Prof. Dr. Seock-Sam –Kim

DEPUTY CHAIRMAN
Dr. Md. Mizanur Rahman
Dr. Mohd. Suffian Misran
Dr. Wan Khairul Muzammil

SECRETARIAT
Dr. Mohd Kamel Wan Ibrahim

EDITORIAL COMMITTEE
Professor Dr. Seock-Sam Kim
Assoc Prof. Dr. Willey Liew Yun Hsien
Dr. Mohd Azlan Bin Ismail
Dr. Md. Mizanur Rahman
Dr. Mohd. Suffian Misran

TREASURER
Dr. Mohd Suffian Misaran

PRINTING AND PUBLISHING
Dr. Mohd Azlan Bin Ismail

INDUSTRY AND INTERNATIONAL AFFAIR
Ir. Mohd Adzrie Bin Radzali
Ir. Mohd Amran bin Hj. Madlan
Ir. Sunderaeswaran Nayar

EVENT MANAGEMENT
Dr. Noor Ajian Mohd Lair (PNU – UMS Workshop Coordinator)
Ir. Mohd Adzrie Bin Radzali (ICMEEAM- 2018 Floor Coordinator)
Dr. Melvin Gan Jet Hong
Mr. Abdullah B. Mohd Tahir
Mr. Azmi Bin Jumat
Mr. Muhin Bin Haji
Ms. Siti Khadizah Binti Yaakub
Ms. Fadzlita Binti Mohd Tamiri
Ms. Adriana Erica Amaludin

IOP Conf. Series: Materials Science and Engineering **606** (2019) 011001 doi:10.1088/1757-899X/606/1/011001

International Conference On Mechanical Engineering, Energy And Advanced Materials (ICMEEAM 2018)
29 – 30 November, 2018, Kota Kinabalu, Sabah Malaysia

UMS
UNIVERSITI MALAYSIA SABAH

FOREWORD FROM THE DEAN
FACULTY OF ENGINEERING, UMS

Welcome to the International Conference on Mechanical Engineering, Energy and Advanced Materials (ICMEEAM) 2018. I would like to thank all participants for your interest and engagement in this conference. The Energy Research Unit (ERU), the Mechanical Engineering Programme and the Materials & Mineral Research Unit (MMRU) of Faculty of Engineering, UMS are delighted to have the opportunity to organize this conference together with the Eco-Campus Management Centre, UMS; Pusan National University (PNU), South Korea and Kyungpook National University (KNU), South Korea.

The field of Mechanical Engineering, Energy and Advanced Materials has been a great contribution to technology and human progress in numerous ways. The goal of this event is to bring together the technologies and researchers who share interest in this field of areas and hopes to create opportunities for both the research and industrial communities to meet, discuss and share their findings and outcomes essential for progressing through the challenges of the 21st Century. The conference also aims to provide an international environment with distinguished speakers from different countries for the participants throughout the world to exchange their views and share their ideas in a vast field of mechanical engineering, energy and advanced material science.

Therefore, I strongly hope the outcomes of this conference will stimulate and accelerate the advancement in knowledge and technologies, and increase the synergy between academic and industry professionals working in this area with emerging networks, applications and services, and to exchange ideas and explore new avenues of collaborations.

Finally, I would like to acknowledge the support and effort of the various parties involved to their patience and perseverance in ensuring this conference a successful event.

Thank you very much.

PROFESSOR IR. DR. ABDUL KARIM BIN MIRASA
Dean
Faculty of Engineering
Universiti Malaysia Sabah

IOP Conf. Series: Materials Science and Engineering **606** (2019) 011001 doi:10.1088/1757-899X/606/1/011001

International Conference On Mechanical Engineering, Energy And Advanced Materials (ICMEEAM 2018)

29 – 30 November, 2018, Kota Kinabalu, Sabah Malaysia

UMS
UNIVERSITI MALAYSIA SABAH

FOREWORD FROM THE CONFERENCE CHAIRMAN

It is my great pleasure to welcome you to the International Conference on Mechanical Engineering, Energy and Advanced Materials (ICMEEAM) 2018, which takes place in beautiful coastal city Kota Kinabalu, Sabah, Malaysia on the 29th to 30th November 2018. Sabah is well known for its rich in cultural, biodiversity, landscape and resources. I hope you will take this great opportunity to explore Sabah while you are here.

It has been a real honor and privilege to serve as the Chairman of this conference. The Energy Research Unit (ERU), Mechanical Engineering Programme (HK08) and the Materials & Mineral Research Unit (MMRU) of Faculty of Engineering, Universiti Malaysia Sabah (UMS) have taken this opportunity to organize this conference in collaboration with Eco-Campus Management Centre, UMS; Pusan National University (PNU), South Korea and Kyungpook National University (KNU), South Korea. ICMEEAM 2018 is also honored to have five distinguished speakers, Dr. Dae-Eun Kim from the Yonsei University, South Korea, Dr. Yusri Yusof from Universiti Tun Hussein Onn Malaysia (UTHM), Malaysia, Prof. Dong Gyu Ahn from Chosun University in Gwang-Ju, South Korea, Prof. Dr. Hussain H. Al-Kayiem from Universiti Teknologi PETRONAS (UTP), Malaysia and Dr. Jun Ho Choi from University of Tokyo, to share their vision on current research topics during the conference.

Thanks to the effort of the ICMEEAM 2018 committee members, we have received 70 paper submissions from 10 countries. All of the submissions were reviewed by the conference technical committee members. All accepted papers will be appeared in IOP Conference Series: Material Science and Engineering (Abstracting and indexing in ISI Web of Science (Conference Proceedings Citation Index), Scopus, Ei Compendex and among others. Selected papers will be published in one of these SCOPUS indexed journals: Journal of Mechanical Engineering Research & Developments (JMERD). Papers related to university campus sustainability can be included in Springer's upcoming book, 'Green Engineering for Campus Sustainability.

The conference would not have been possible without the enthusiastic and diligent effort from all the conference organizers, committee members, advisors and reviewers. We are also grateful to all the authors who trusted the conference with their work. They are the ones who help this conference possible through their hard work in preparation of the manuscripts. On behalf of the organizing and ICMEEAM 2018 committees, I welcome you all to Sabah and we sincerely hope that you enjoy ICMEEAM 2018 and your stay in Kota Kinabalu, Sabah.

Prof. Dr. Seock-Sam Kim
ICMEEAM 2018 Chairman

IOP Conf. Series: Materials Science and Engineering **606** (2019) 011002 doi:10.1088/1757-899X/606/1/011002

Peer review statement

All papers published in this volume of *IOP Conference Series: Materials Science and Engineering* have been peer reviewed through processes administered by the proceedings Editors. Reviews were conducted by expert referees to the professional and scientific standards expected of a proceedings journal published by IOP Publishing.

IOP Conf. Series: Materials Science and Engineering **606** (2019) 012001 doi:10.1088/1757-899X/606/1/012001

Optimisation of the Hardness AZ31B Reinforced with Lead and Carbon Nanotubes using the Response Surface Method

M F Abdullah[1]*, S Abdullah[2], M S Risby[1], M Z Omar[2] and Z Sajuri[2]
[1]Department of Mechanical Engineering, Faculty of Engineering, Universiti Pertahanan Nasional Malaysia, Kem Sg. Besi 57000 Kuala Lumpur, Malaysia
[2] Centre for Materials Engineering and Smart Manufacturing, Faculty of Engineering & Built Environment, Universiti Kebangsaan Malaysia, 43600 UKM Bangi, Selangor, Malaysia

Abstract. This paper presents the optimisation of the hardness of the material, AZ31B, reinforced with a percentage of lead and carbon nanotubes (CNTs). Currently, the new automotive era is focusing on lighter and stronger materials. Therefore, in this study, although magnesium alloy showed characteristics of high strength, yet it still needed to be enhanced in terms of strength and elongation. Hence, AZ31B was selected to enhance the strength through the addition of lead and CNT reinforcement particles. The Response Surface Method (RSM) was used to optimise the percentage composition of lead and CNT. From the RSM, a mathematical model can be obtained for the optimization of the given parameter. The RSM method was validated through an experimental analysis using the Disintegrated Melt Deposition (DMD) technique for the sample preparation. From the results, it was shown that the optimised composition due to the addition of lead was between 1% and 5%, while with CNT it was between 0.1% and 0.5%. The suggested reinforcement materials increased the material hardness of the AZ31B material by up to 27%. Hence, the proposed RSM was suitable for the optimization of the reinforcement materials.

1. Introduction

The demand for lightweight materials for automotive applications is very high because these materials can reduce the load of the vehicle and, at the same time, save on fuel costs [1]. One of the most important material properties in automotive applications is the strength, which allows the device to withstand the effects of various conditions. The strength of this material is also directly proportional to the hardness of the material [2]. Therefore, the development of such a composite was switched to that of an alloy composite, namely magnesium alloy, as the lightweight material. Magnesium alloy is used mostly in the automotive and aerospace industry and falls within the AZ31B alloy series, which provide suitable strength properties for various applications [3].

Magnesium has a hexagonal close-packed structure (HCP). This structure affects the strength and ductility of materials because there are voids between the molecules [4]. Therefore, it is necessary to have a material that is able to fill these voids in order to prevent the structural collapse of the material. The most suitable materials to fill those voids are carbon nanotubes (CNTs). CNTs have unique characteristics that can generate strong materials with enhanced energy absorption properties [5], and many researchers have acknowledged that CNTs can improve the strength of materials. CNT materials are produced from nano-technology for the purpose of filling the voids in structures and producing van der Waals bonds within them [6]. In addition, lead (Pb) materials are also being used extensively to increase the absorption of energy as well as to enhance the ductility of materials. Since an increase in the ductility of a material is directly proportional to the elasticity of the material, an increase in the lead content can have an effect on the enhancement of the ductility of AZ31B [7].

Magnesium alloy, AZ31B, is a material with high-energy absorption. However, this material is weak in terms of its ductility and hardness. The addition of CNT to a material can increase the strength and hardness of that material, while its ductility can be increased by the addition of lead. These additions can improve the material properties of magnesium alloy, making it better than the original material. Therefore, this study was conducted to obtain the optimum hardness by adding a percentage of CNT and lead to AZ31B using the RSM. The addition of an optimum percentage of CNT and lead can increase the hardness of the experimental materials.

2. Methodology

2.1. Sample preparation

The matrix material used was the magnesium alloy, AZ31B, in the form of an ingot. CNT, with an external diameter of 10-30 nm, density of 2.1 g/cm^3 and a purity of more than 90%, was added to the AZ31B. Besides that, the lead powder that was added had a purity of 99.9%, with a density of 11.342 g/cm^3. This process of adding foreign substances into the AZ31B is known as the Disintegrated Melt Deposition (DMD) method. In this case, the AZ31B ingot was heated in an induction furnace until it melted. This heating was conducted in a chamber where the air had been sucked out to create a partial vacuum before argon gas was piped through the combustion chamber at a speed of 25 mm/min [8]. When the ingot had melted, the CNT and lead compositions were channelled into the chamber concerned. The percentage of CNT and lead that was added depended on the experimental design, as shown in Table 1. Figure 1 shows the machine induction furnace where the substances were added to the melted magnesium alloy.

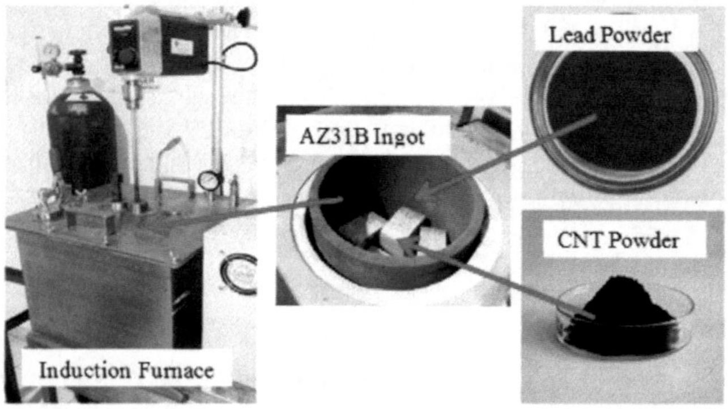

Figure 1: Induction furnace for melting magnesium alloy and reinforcement materials.

2.2. Matrix combination for Response Surface Method

The development of the magnesium alloy, which involves the addition of CNT and lead, uses a statistical method known as the Response Surface Method (RSM). RSM is a practical and economical process that is used by many researchers to obtain the optimum parameters for an experiment [9]. The mathematical model that is generated by the RSM gives the relationship between the response and the given input as equation (1):

$$y = f(x_1, x_2, ..., x_k) + \varepsilon \quad (1)$$

where y is the response, f is the unknown function of the response, x_1, x_2, ..., x_k are the independent variables, also known as the original variables, ε is the statistical error that represents a variable source that has not been taken into account in the function, f. ε is usually regarded as a normal distribution with a min variation of zero.

Table 1 gives the parameters that were tested in the RSM experiment. Five factors were involved, namely the CNT percentage, lead percentage, temperature, endurance duration and mixing speed. All the factors were tested at two levels, that is, the minimum and maximum levels.

IOP Conf. Series: Materials Science and Engineering **606** (2019) 012001 doi:10.1088/1757-899X/606/1/012001

Table 1: Main parameters and their important levels that were affected

Level codes	Maximum	Minimum
Percentage of CNT	0.5	0.1
Percentage of Lead	5.0	0.5
Temperatures (°C)	660	750
Holding time (s)	1800	900
Stirrer speed (rpm)	500	250

The parameters that were obtained were based on an earlier study by Katsuyoshi et al. (2010) [10], which found that a suitable composition of CNT for their study was around 0.5%. Srinivasan et al. (2007) showed that an increase in the lead content by as much as 5% of the original composition of the material would have a positive effect on the energy absorption [7]. The temperature parameter was taken into account based on the melting point of magnesium and on the study by Nguyen et al. (2008), which used a superheated temperature of 750 °C [8]. Other parameters were also discussed by previous researchers.

2.3. Hardness Observation
A hardness test was carried out to determine the change in the hardness of the alloy due to CNT and lead. A Vickers hardness (HV) test was conducted using a power scale of 300 gf in reference to the ASTM E384 standard [11]. Figure 2 shows the hardness testing machine that was used in this experiment.

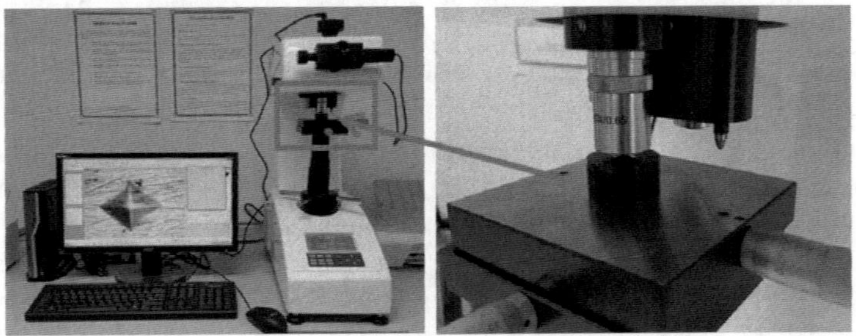

(a) (b)

Figure 2: (a) Vickers tester used for hardness characterization, (b) view of placed sample

3. Results and discussion
Table 2 shows the number of runs obtained in the RSM with regard to the response to the hardness of the material. There were 25 runs altogether, which gave different hardness results. This RSM table used a model with one midpoint and three replications to get a significant comparison in the statistical analysis. Figure 3 shows a comparison of the percentage increase and decrease in the hardness of each material with the parameters as set out in Table 2 against that of the original material. The original hardness of the AZ31B was 57.7 HV for a scale of 300 gf. From the hardness test conducted, some materials experienced a reduction in hardness, while others experienced an increase. It was observed that the highest reduction in hardness was at run 7, i.e. 46.7 with a reduction of -19.1%. The highest increase in hardness was at run 22, i.e. 72.9 with an increase of 26.3% compared to the original material. The median of increase in hardness was at run 12, i.e. 62.0 with an increase of 7.5% compared to the original material. According to the DOE table that was obtained, run 8 was the run at the midpoint of the model. Run 8 had a hardness of 65.5, i.e. 13.5% higher than that of the original material, AZ31B.

Table 2: Number of DOE Runs for RSM

Run Order	Hardness	Run Order	Hardness
1	57.2	14	55.0
2	56.6	15	68.1
3	65.7	16	63.3
4	60.2	17	70.0
5	60.2	18	60.1
6	68.2	19	65.3
7	46.7	20	53.8
8	65.5	21	61.7
9	66.2	22	72.9
10	55.4	23	70.4
11	67.7	24	64.4
12	62.0	25	65.8
13	66.0		

Figure 4 shows the normal probability graph for the hardness of each AZ31B composite with a combination of CNT and lead. This normal probability uses the P-value as its reference point. The P-value taken for this experiment was 0.001. The P-value is important for determining whether the data has a normal distribution or not. From the data taken for 25 runs, the P-value for hardness was 0.025. The P values obtained for the hardness was much higher than the P-value reference point, thus indicating that the hardness data for the magnesium alloy had a normal distribution. The hardness of the magnesium alloy had an average value, μ, of 63.87 HV and a standard deviation, σ of 5.088. From Figure 4, it can be seen that the hardness data for magnesium alloy was located in the border area distribution with a 95% confidence level. A normal distribution was important at the start of the statistical analysis to ensure that the data that was obtained fulfilled the statistical conditions for the next stage of the analysis, i.e. the RSM analysis.

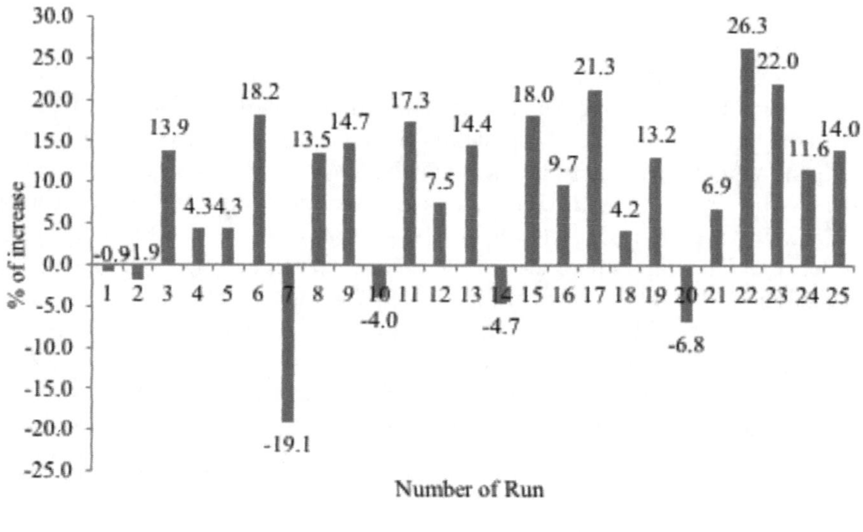

Figure 3: Percentage increase in hardness compared to AZ31B

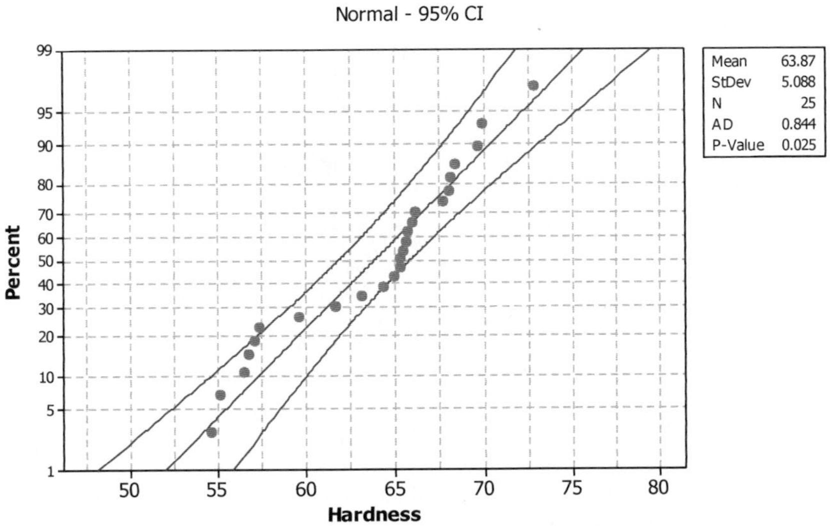

Figure 4: Normal hardness probability graph for each AZ31B composite containing CNT and lead

In the next analysis, the hardness data for the material was employed for the optimization using the RSM. The RSM is a practical, economical and relatively easy method, and many researchers utilize it for the modelling process. It is also successfully used for applications in composite material science [12]. The calculated data was used with a regression method to construct a mathematical model. This mathematical model was taken as the objective function and could be optimized using any mathematical or analytical approach to obtain the appropriate process parameters.

Figure 5 shows the contour plot for the effects of hardness on the percentage composition of CNT and lead in AZ31B. It can be seen that the hardness of the material increased with the addition of CNT in the initial stage, but started to decrease when it exceeded the optimum content for CNT. The percentage increase in lead, on the other hand, did not have a significant impact, as the change that occurred in the hardness was not extreme. This response towards hardness was in the form of a curve, which can be clearly seen in Figure 6, which depicts the plot of the surface hardness towards the percentage increase in CNT and lead. These contours and surface plots took into consideration the midpoint factors that were given, namely a temperature of 705 °C, duration of 22.5 minutes and rotational speed of 375 rpm.

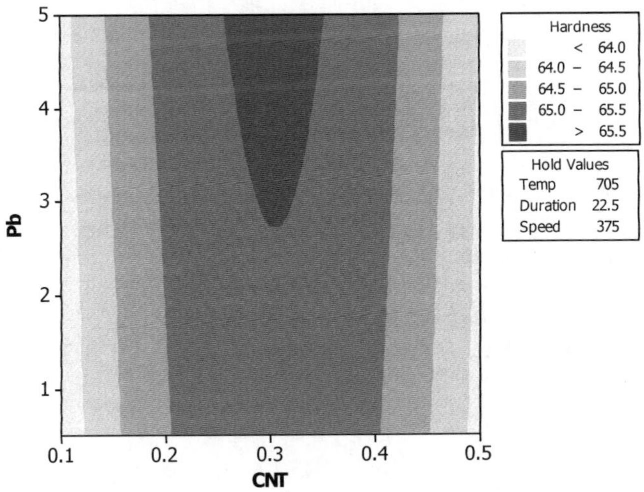

Figure 5: Contour plot for hardness in relation to percentage increase in CNT and lead

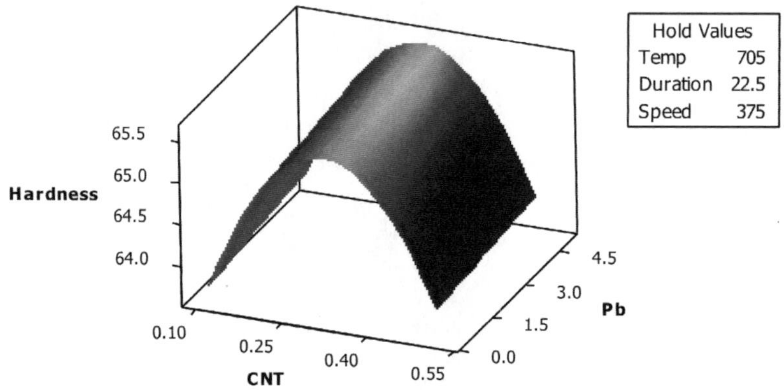

Figure 6: Surface plot for hardness in relation to percentage increase in CNT and lead

From the estimated regression coefficient for hardness and based on the factors given in this experiment, the regression model equation (2) can be derived as follows:

$$\begin{aligned} Hardness = & -32.529 + 25.917(CNT) + 11.769(Pb) + 0.118(Temp) \\ & + 0.328(Duration) + 0.009(Speed) - 42.500(CNT * CNT) \\ & - 0.014(Pb * Temp) - 0.005(Pb * Speed) \end{aligned} \tag{2}$$

The composition of this material could also be optimised with regard to the response of the material using the RSM. Figure 7 shows the optimum percentage content of CNT and lead in the AZ31B composite. It was learned that the optimal content for this response was 0.306 % CNT and 0.5 % lead at a temperature of 750 °C, duration of 30 minutes and rotational speed of 500 rpm. This provided the optimal hardness of 73.70 HV, which was 27.7 % higher than the hardness of the original material, AZ31B.

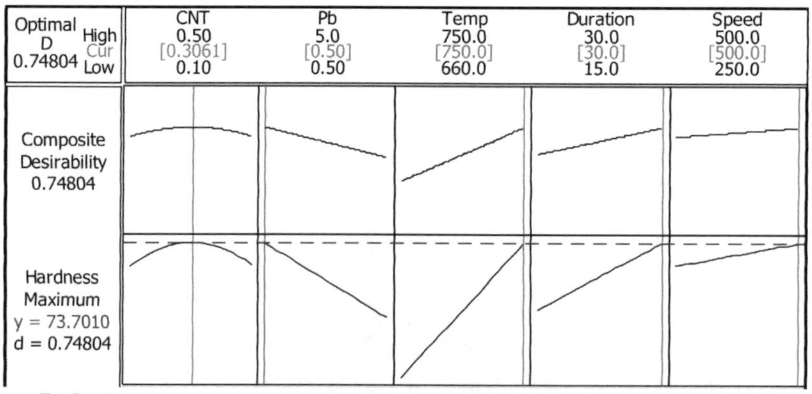

Figure 7: Optimum percentage content of CNT and lead in the AZ31B composite

Figure 8 shows the correlation between the experiment and the simulation for the material hardness. The correlation plot showed that a very high R^2 value of 0.9054 was obtained, which approached the ideal precision value of 1.0. This showed that the experiment generated a hardness equation that could be used in this study.

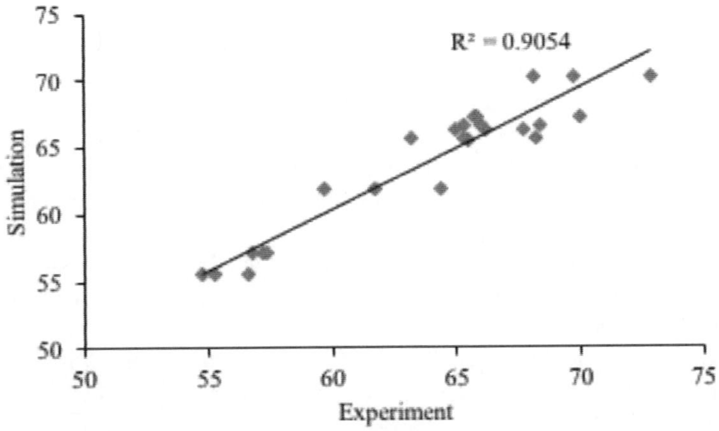

Figure 8: Correlation graph simulation versus experiment for hardness

4. Conclusions

The increase in the percentage of CNT and lead can enhance the hardness of the magnesium alloy, AZ31B. These additions took into account other side factors such as temperature, the melt duration and the rotational velocity during the blending. These factors resulted in a combined hardness from the percentage of CNT and lead. These factors indirectly contributed to the change in the nature of the material being studied. The use of the RSM statistical analysis method enabled the optimum value for the percentage increase in the composition to be known. The optimum composition for CNT and lead was 0.306% and 0.5 %, respectively, which served to increase the hardness by as much as 27.7 % compared to the original material, AZ31B. The hardness that was obtained enhanced the properties of the AZ31B magnesium alloy. This resultant increase in hardness will extend the use of the magnesium alloy, giving rise to a higher level of energy absorption by the magnesium alloy upon impact at high speeds.

Acknowledgments

The authors would like to express their gratitude to Ministry of Higher Education Malaysia via Universiti Kebangsaan Malaysia and Universiti Pertahanan National Malaysia (Research funding: LRGS/2013/UPNM-UKM/DS/04) for supporting this research.

References

[1] Bakshi S R, Lahiri D and Agarwal A 2010 Carbon nanotube reinforced metal matrix composites a review. *International Materials Reviews* vol **55** (1). Pp 41-64

[2] Saad A, Alsubaie, Yi Huang Terence and Langdon G 2017 Hardness evolution of AZ80 magnesium alloy processed by HPT at different temperatures. *Journal of Materials Research and Technology* **6** (4): 378-384

[3] Watari H, Davey K, Rasgado M T and Izawa S 2004 Semi-solid manufacturing process of magnesium alloy by twin-roll casting, *J. Material Processing Technology*, Vol **155-156**: 1662-1667

[4] Jena P K, Kumar K S, Krishna V R, Singh A K and Bhat T B 2008 Studies on the role of microstructure on performance of a high strength armour steel. *Engineering Failure Analysis* **15** 1088–1096

[5] Muhammad R, Mahesh H, Shaik Z, Uday V, Arefin T, Ashok K, Jonathan T and Shaik J 2013 Effects of amino-functionalized MWCNTs on ballistic impact performance of E-glass/epoxy composites using a spherical projectile. *International Journal of Impact Engineering* **57** 108-118.

IOP Conf. Series: Materials Science and Engineering **606** (2019) 012001 doi:10.1088/1757-899X/606/1/012001

[6] Selvamani S T, Premkumar S, Vigneshwara M, Hariprasath P and Palanikumard K 2017 Influence of carbon nano tubes on mechanical, metallurgical and tribological behavior of magnesium nanocomposites *Journal of Magnesium and Alloys* **5** (3): 326-335
[7] Srinivasan A, Pillai U T S and Pai B C 2007 Effect of Pb addition on ageing behavior of AZ91 Magnesium alloy *Material Science and Engineering* **452-453** p87-92
[8] Nguyen Q B and Gupta M 2008 Increasing significantly the failure strain and work of fracture of solidification processed AZ31B using nano-Al2O3 particulates *Journal of Alloys and Compounds* **459**(1–2), p244–250
[9] Shokuhfar A, Khalili S M R, Ghasemi A F, Malekzadeh K and Raissi S 2008 Analysis and optimization of smart hybrid composite plates subjected to low-velocity impact using the response surface methodology (RSM) Thin-Walled Structures **46**: 1204 – 1212
[10] Katsuyoshi K, Hiroyuki F, Junko U, Hisashi I, Bunshi F and Morinobu E 2010 Microstructural and mechanical analysis of carbon nanotube reinforced magnesium alloy powder composites *Materials Science and Engineering* A **527** : 4103–4108
[11] ASTM E384-2011 *standard test method for knoop and vickers hardness of materials* P 43
[12] Shokuhfar A, Farsani R, Raissi S and Sedghi A 2007 Optimization of carbon fibers made up of commercial polyacrylonitrile fibers using screening design method. *Mater Sci: An Interdiscip J Phys, Chem Technol. Mater* **25**(1):113–20

IOP Conf. Series: Materials Science and Engineering **606** (2019) 012002 doi:10.1088/1757-899X/606/1/012002

Characterization of carbonized waste materials: Rice husk and saw dust

G J H Melvin[1,2*], Kah Fei Chai[2], Fadzlita Mohd Tamiri[2]

[1] Material and Mineral Research Unit (MMRU), Universiti Malaysia Sabah, Jalan UMS, 88400 Kota Kinabalu, Sabah, Malaysia

[2] Faculty of Engineering, Universiti Malaysia Sabah, Jalan UMS, 88400 Kota Kinabalu, Sabah, Malaysia.

*E-mail: melvin.gan@ums.edu.my

Abstract. Waste materials can be categorized into household, commercial, agricultural, synthetic wastes, and so on. Agricultural wastes play a great role as an alternative material to be derived into valuable carbon materials, due to their low cost and abundancy, in contrast to conventional carbon material sources. Feasible carbonization process of agricultural wastes can be conducted by using heat treatment. The focused agricultural wastes in this study are rice husks (RH) and saw dusts (SD). They are converted into carbon materials through heat treatment under the presence of inert gas at 500°C and 800°C for 2 hours. Then, their characteristics such as structure, morphology, elemental composition, and crystallinity are investigated by utilizing scanning electron microscopy (SEM), Energy Dispersive X-ray (EDX) spectroscopy, and X-ray Diffraction (XRD). Similarly, the carbonized RHs and SDs can be classified as amorphous carbon and possess porous structure. The characteristics of carbonized RHs and SDs indicate that they could be a promising material to be utilized in particular or various applications.

1. Introduction

Rice husk (RH) is an agro-based lignocellulosic waste material that is abundant in various countries, majorly produced from by-product of rice milling industries [1, 2]. Based on the report from Food and Agriculture Organization of the United Nations (FAO), in year 2016, world rice production is approximately 741 million tonnes, and approximately 2.3 million tonnes was contributed by Malaysia. RH is accountable for around 20% from the rice production, which is ready to be utilized [3]. Generally, RH is manipulated as low value energy resource, discarded, or simply burnt at the field, which will influence the air quality and threaten the environment [4]

On the other hand, saw dust (SD) is an ample lignocellulosic waste material obtained from the processing of wood including cutting, grinding, drilling, scraping, or pummeling wood with saw or any other cutting tool used in sawmills, factories, or in the household activities [5-7]. Based on the report from FAO, in year 2017, world wood residue production is approximately 225 million m^3, and approximately 1.7 million m^3 was produced in Malaysia. SD, together with wood offcuts and chips, is accountable for around 20% of the total input mass [6]. Commonly, SD is accumulated or combusted in factories or mills, which hinder its effective utilization and unfavorable to the environment [5, 6].

Therefore, it is essential to conduct research on RH and SD in order to fully utilize and unlock their potential for broad applications. Both RH and SD are lignocellulosic materials, which is in favor to

IOP Conf. Series: Materials Science and Engineering **606** (2019) 012002 doi:10.1088/1757-899X/606/1/012002

convert them into raw carbon materials [8]. The conversion simply can be accomplished through carbonization process using heat treatment [9]. Due to the low cost and abundancy of RH and SD, in contrast with other carbon materials such as carbon nanotubes, carbon black, graphene, and so on, they can be expected to play a significant role to be fully utilized as the source of carbon materials. In this study, the RH and SD were carbonized and then their structure, morphology, elemental composition and crystallinity were characterized and compared.

2. Experimental

Dried RHs and SDs without further modification were used as the source to be carbonized. They were carbonized at different temperatures of 500°C and 800°C. The raw RHs and SDs were carbonized at 500°C, 2 h using furnace, under the presence of argon gas. Similar step was repeated for carbonization at 800°C. All samples were cooled down naturally and then grounded into powder before further characterization. RHs and SDs carbonized at 500°C will be denoted as RH500 and SD500; RHs and SDs carbonized at 800°C will be denoted as RH800 and SD800, hereafter.

The samples were observed using scanning electron microscopy (SEM; Hitachi S-3400N) equipped with energy dispersive X-ray (EDX), in order to investigate the structure and morphology. EDX was utilized to characterize the elemental composition. X-ray diffraction (XRD; Philips X'Pert) measurement was performed in order to determine the crystallinity of produced carbon materials.

3. Results and discussions

The mass difference of the samples before and after carbonization at different temperatures is tabulated in table 1. As the carbonization temperature increased, the mass difference also increased. The decomposition of volatile materials during carbonization process leads to the mass loss. Overall, SD showed higher mass loss when compared to RH, for both temperatures. The highest mass difference was marked by SD800, which is 77%.

Table 1. Mass difference after carbonization process

Material	Mass Before (g)	Mass After (g)	Mass Difference (%)
RH500	30.0	12.7	58
RH800	30.0	11.4	62
SD500	30.0	8.3	72
SD800	30.0	6.8	77

SEM images of carbonized RH at different temperatures are depicted in figure 1. Porous structure is obvious, and can be clearly seen from RH500 and RH800. The pores can be considered as macroporous, in μm range. No drastic structure changes are observed at this temperature range, dissimilar to Ref. [9]. From figure 1(b), sand-like particles (red circled) can be observed. Relatively, higher silicon is traced via EDX at these particles, which indicated they are rich in silica. SEM images of carbonized SD at different temperatures are depicted in figure 2. Similarly, porous structures can be observed from SD500 and SD800. From figure 2(b), smaller pores, around nm range, can be observed. However, they are not evident for SD800. The porous structure of carbonized RH and SD can provide good adhesion with matrix materials such as polymer matrix, which can acts as reinforcement.

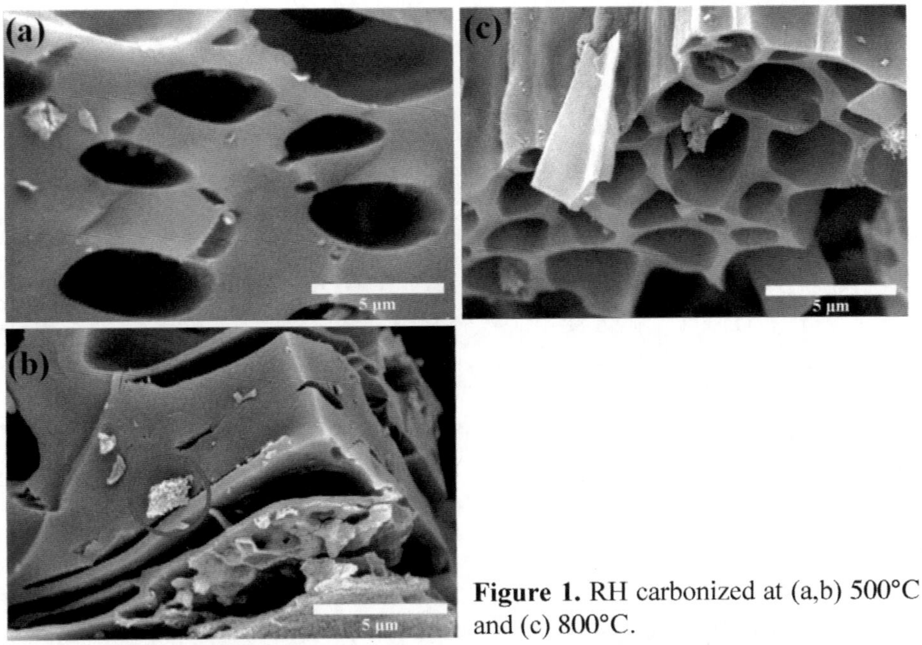

Figure 1. RH carbonized at (a,b) 500°C and (c) 800°C.

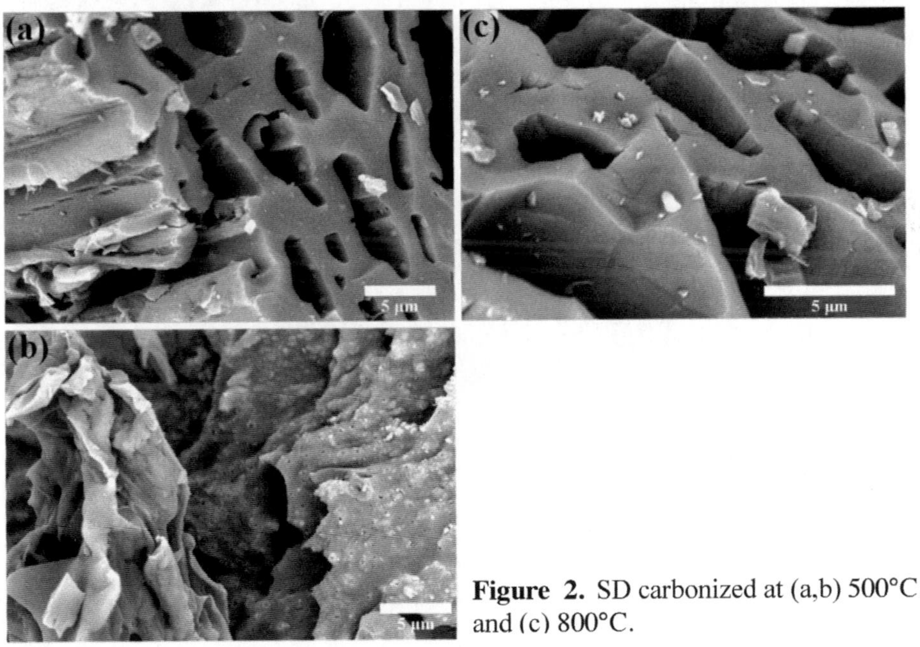

Figure 2. SD carbonized at (a,b) 500°C and (c) 800°C.

EDX spectroscopy measurement was conducted to determine the elemental composition of carbonized RH and SD, and summarized in table 2. The main component obtained after carbonization process is carbon. In contrast with carbonized SD samples, silicon element was only observed in carbonized RH samples. The silica absorption of rice plants in the form of silicic acid from the soil is thinkable [10]. Furthermore, oxygen also can be traced, and this can be linked with certain amounts of functional groups such as hydroxyl and carboxyl [11, 12]. Other elements can be referred to impurities, mineral absorbed by paddy or wood, fertilizers, and so on. Overall, carbonized SD produced higher carbon purity, around 85%, in contrast to carbonized RH. This is in accordance to the higher mass

IOP Conf. Series: Materials Science and Engineering **606** (2019) 012002 doi:10.1088/1757-899X/606/1/012002

difference (table 1) of SD. It is also worth to notice that the increment of carbonization temperature from 500°C to 800°C for SD did not result in higher carbon contents.

Table 2. Elemental composition of carbonized RH and SD.

Material	Element (wt.%)			
	Carbon	Oxygen	Silicon	Others
RH500	73.42	9.95	4.05	12.58
RH800	78.74	11.50	1.62	8.15
SD500	85.38	7.34	-	7.28
SD800	85.43	5.52	-	9.05

XRD result of RH500 is illustrated in figure 3. Generally, all carbonized RH and SD showed similar XRD patterns with RH500. This further revealed that both carbonized RH and SD possess typical amorphous carbon broad peak (002), at around $2\theta = 22\sim23°$. In order to further characterize the results, fitting line is fixed to the main peak. The results obtained from fitting line are tabulated in table 3. Based on the fitting line of carbon peak for carbonized RH and SD, the full width at half maximum (FWHM) is around 8~9°. This further confirms that the carbonized RH and SD are similar, regardless the carbonization temperature, which they can be classified as low crystallinity carbon materials.

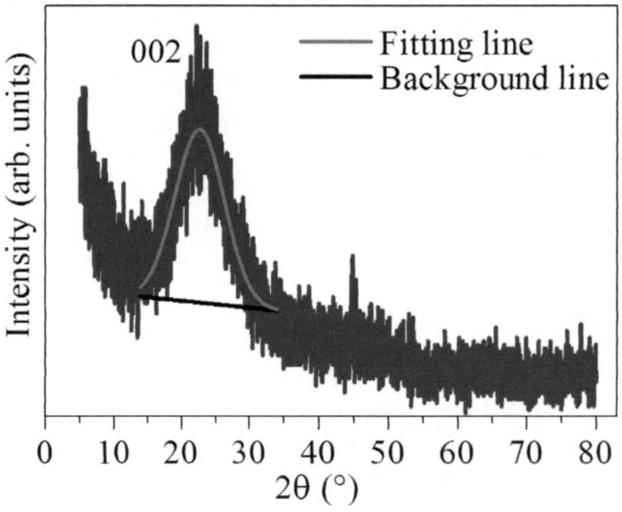

Figure 3. XRD of RH500

Table 3. XRD results based on fitting line

Material	Peak Position (°)	FWHM (°)
RH500	22.662	8.532
RH800	23.388	8.138
SD500	22.533	9.386
SD800	23.545	8.930

Notably, this type of carbon materials usually further developed as the activated carbon through physical or chemical activation for applications such as absorbent, energy storage devices, and so on [3, 4, 6, 13]. Furthermore, they also exhibited high potential to be developed as the multifunctional composite materials for various applications such as hybrid nanomaterials, electromagnetic wave absorber, and so on [14-16]. They can be exploited as an abundant and cost effective material for various applications.

4. Conclusions

Carbon materials derived from RHs and SDs were prepared by carbonizing them at 500°C and 800°C. Their structure, morphology, elemental composition and crystallinity were evaluated through SEM observation, EDX evaluation, and XRD measurement. Both carbonized RHs and SDs showed porous structure and significantly high content of carbon element. Furthermore, they can be categorized as low crystallinity material which is amorphous carbon. It is undeniable that their performance might be limited in comparison to the high grade or high crystallinity carbon materials. However, since they are available in large scale, greener technology to produce bulk quantity of valuable carbon materials from agricultural wastes can be promoted.

References

[1] Christopher F, Bolatito A and Ahmed S 2017 *Int. J. Sustainable Built Environ.***6** 675-92
[2] Arora S, Kumar M and Kumar M 2012 *J. Reinf. Plast. Compos.* **31(2)** 85-93
[3] Guo Y, Yu K, Wang Z and Xu H 2000 *Carbon* **41** 1645-87
[4] Chen Y, Zhu Y, Wang Z, Li Y, Wang L, Ding L, Gao X, Ma Y and Guo Y 2011 *Adv. Colloid Interface Sci.* **163** 39-52
[5] Ahmed W, Khushnood R A, Memon S A, Ahmad S, Baloch W L and Usman M 2018 *J. Cleaner Prod.* **184** 1016-27
[6] Foo K Y and Hameed B H 2012 *Bioresour. Technol.* **111** 425-32
[7] Shukla A, Zhang Y-H, Dubey P, Margrave J L and Shukla S S 2002 *J. Hazard. Mater.* **95** 137-52
[8] Wang L, Guo Y, Zhu Y, Li Y, Qu Y, Rong C, Ma X and Wang Z 2010 *Bioresource Technol.* **101** 9807-10
[9] Melvin G J H, Wang Z, Siambun N J and Rahman M M 2017 *IOP Conf. Ser.: Mater. Sci. Eng.* **217** 012017
[10] Liu N, Huo K, McDowell M T, Zhao J and Cui Y 2013 *Sci. Rep.* **3** 1919
[11] Melvin G J H, Ni Q-Q and Natsuki T 2016 *Polym. Compos.* **37(1)** 262-9
[12] Melvin G J H, Ni Q-Q and Natsuki T 2014 *Appl. Phys. A* **117(4)** 2043-50
[13] Muramatsu H, Kim Y A, Yang K S, Cruz-Silva R, Toda I, Yamada T, Terrones M, Endo M, Hayashi T and Saitoh H 2014 *Small* **10(14)** 2766-70
[14] Melvin G J H, Wang Z, Ni Q-Q, Siambun N J and Rahman M M 2017 *AIP Conf. Proc.* **1901** 020002
[15] Melvin G J H, Wang Z, Ni Q-Q, Siambun N J and Rahman M M 2017 *IOP Conf. Ser.: Mater. Sci. Eng.* **229** 012024
[16] Ismail H, Nasaruddin M N and Rozman H D 1999 *Eur. Polym. J.* **35(8)** 1429-37

IOP Conf. Series: Materials Science and Engineering **606** (2019) 012003 doi:10.1088/1757-899X/606/1/012003

Adsorption of Ammonium Ion Using Zeolite, Chitosan, Bleached Fibre and Activated Carbon

N N Safie, A Y Zahrim*, M Rajin, N M Ismail, S Saalah, S M Anisuzzaman, A D Rahayu, H Huslyzam, R Jennisha, T T H Calvin
Faculty of Engineering, Universiti Malaysia Sabah, Jalan UMS, 88400 Kota Kinabalu

E-mail: zahrim@ums.edu.my

Abstract. Several materials have been studied for the adsorption of ammonium ion from synthetic solution. Zeolite is having the highest adsorption capacity (3.160 mg/g) with fast reaction time followed by chitosan (2.5770 mg/g). Activated carbon derived from Tarap (*Artocarpus odoratissimus*) and rice husk shows better performance than the low cost bleached fibres.

1. Introduction

Ammonium is the most basic form of nitrogen and commonly present in the waterways due to the discharge from urban runoff, industrial discharges and untreated sewage discharge[1]. Several pollution cases due to NH_4^+ has been recorded in Malaysia river [1] indicating the needs for proper treatment. To minimise the discharging of ammonium ion into the river, gross pollutant trap can be coupled with specific adsorbent for ammonium ion removal. Agricultural waste has been used as low-cost bio sorbents since it is low cost, renewable, wide range of sources as well as contain functional groups such as carboxyl group and hydroxyl group that bind with contaminants during adsorption [2]. *Langsat* peels, jackfruit seeds and *moringa* peels have been used in NH_3-N removal[3]. Other than that, activated carbon has been used widely in adsorption especially activated carbon derives from low-cost materials such as animal bones[4] and waste bamboo[5] which are also known to have highly developed surface area, porosity and large adsorption capacity.

Biopolymer such as chitosan has being used because it is natural, biodegradable, reusable and non-toxic polysaccharide from crustacean shells such as shrimps and crabs [6]. Fibers are known to be rich in lignocellulosic materials that comprised of three main components namely lignin, cellulose and hemicellulose that contain functional groups responsible for the adsorption process to occur [7]. In this study, zeolite, chitosan, bleached fibres from coconut coir, water hyacinth, empty fruit bunch (EFB) and activated carbon from rice husk and tarap *(Artocarpus odoratissimus)* skin have been tested to remove ammonium ion by adsorption.

2. Materials and methods

Zeolites and high molecular weight chitosan were purchased commercially. Banana stems, coconut coir, water hyacinth, *tarap* skin and rice husk were obtained from Kota Kinabalu, Salut Commercial Centre, Tuaran river, Papar and Kilang Padi Sri Keranahan, Kota Belud respectively. Activated carbon from rice husk and tarap skin were prepared according to the method by Binanipuram and Kerala [8] and Wang et al. [9] respectively. Zeolite, high molecular weight chitosan, rice husk activated carbon, tarap skin activated carbon, bleached EFB, bleached water hyacinth, bleached coconut coir and bleached banana stem are denoted as zeolite, HMWC, RHAC, TSAC, EFB, WH, CC and BS respectively.

The NH_4Cl stock solution was prepared from a 99.5% pure anhydrous solid by dissolution in distilled water. NH_4Cl solution with a concentration of 0.05 mg/g was used for the batch adsorption experiment. The samples were then analysed using it was analysed using Jasco UV-vis 650 Bio-spectroscopy with maximum wavelength of 425 nm. Prior to analysis, Nessler reagent was used for NH_3-N compound detection by following the standard USEPA Nessler Method No.8038. The concentration of ammonium ions adsorbed at equilibrium was calculated using Equation (1).

$$q_e = \frac{(C_o - C_e)V}{W} \quad (1)$$

where q_e (mg/g), Co (mg/L), C_e(mg/L), V (dm^3) and W (g) are the concentration of ammonium ions adsorbed at equilibrium, liquid-phase concentrations of initial adsorbate and equilibrium, volume of solution and mass of activated carbon used, respectively.

3. Results and Discussions

As shown in Figure 1 is the concentration of NH_3-N being adsorbed by BS, EFB, WH, CC, zeolite and CC against time in minute. Based on the results summarized in Table 1, zeolite has the highest adsorption capacity followed by HMWC, RHAC, TSAC, BS, EFB, CC and the lowest was WH with values of 3.160 mg/g, 2.577 mg/g, 2.269 mg/g, 0.409 mg/g, 0.317 mg/g, 0.258 mg/g and 0.018 mg/g respectively. Zeolite has reached equilibrium in 2 minutes. Research regarding the sorption kinetics studies that the ammonium ion removal using zeolite occurred very rapidly and equilibrium was reached within few minutes ranging between 10 min and 90 min[10][11]. High selectivity and reasonable adsorption capacity of zeolite can be attributed to their unique structural chemistry, for instance, Si/Al ratio, pore size and high specific surface area [12]. Natural zeolite used by Beebe et al. [13] recorded adsorption capacity of 5.02 mg/g but required longer time to reach equilibrium contact time. This behaviour may be ascribed to the quick utilization of the most readily available adsorbing sites of the zeolite used in this work that has contributed to fast diffusion and attainment of rapid equilibrium[10]. Also, the CEC of zeolite used in Beebe et al. [13] was higher 180–195 meq/100 g compared to CEC zeolite used in this work, 140 meq/100 g which explains that it can absorbs more cation (NH_4^+). MCW has higher adsorption capacity compared to bleached fibers and activated carbon and reached equilibrium in 1 minute. HMCW has high deacetylation degree and carries more positive charges that exhibit better performance in removing pollutants[14].

Activated carbon from rice husk and tarap recorded higher adsorption capacity compared to the bleached fibers but with longer time to reach equilibrium. The chemical and activation process has enhanced its porous structure and surface area which attributes to higher adsorption capacity as compared to raw bleached fibers[15]. TSAC has lower adsorption capacity since higher activation temperature was used, 800 °C compared to rice husk 300°C. The lower temperature biochar described by Gaskin *et al.*[16] was shown to have a higher degree of oxygen surface functional groups. The content of silica in most rice husk may greatly enhance the surface area of rice husk compared to tarap skin[17]. Bleached fibers used in this investigation namely BS, EFB, WH and CC with adsorption capacity 0.4085 mg/g, 0.3171 mg/g, 0, 0176 mg/g and 0.2577 mg/g respectively. Bleached fibers recorded the lowest NH_3-N adsorption capacity compared to activated carbon, zeolite and HMWC. Nevertheless, unbleached EFB done by Zahrim et al. [18] recorded higher adsorption capacity, 0.828 mg/g at the same initial concentration and 94-97 % NH_3-N removal when used as the medium in filtration [19].

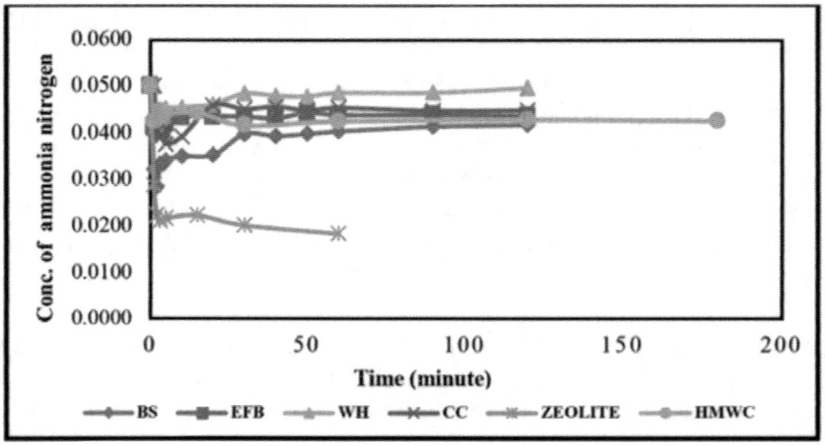

Figure 1. Adsorption of ammonia nitrogen against time using BS, EFB, WH, CC, zeolite and HMWC

Table 1. Total time to reach equilibrium, adsorption capacity and comparison with other authors for zeolite, HMWC, BS, EFB, WH, CC, TSAC and RHAC

Type (s) of adsorbent		Equilibrium contact time (min)	Adsorption capacity (mg/g)	Initial conc. (mg/L)	Ref.
Zeolite		2	3.160	50	This study
		60	1–6	100-500	[20]
		-	2.8	200	[21]
		2880	5.03	40	[13]
		1440	1.9	10	[22]
HMWC		1	2.5770	50	This study
		180	6.933	7.35	[6]
Bleached fibers	BS	30	0.4085	50	This study
	EFB	10	0.3171	50	This study
		40	0.828	50	[18]
	WH	30	0.0176	50	This study
	CC	20	0.2577	50	This study
Activated Carbon	TSAC	20	1.8750	50	This study
	RHAC	20	1.8154	50	This study
		24	0.87 – 2.20	5	[17]

4. Conclusion

The adsorption capacity of zeolite, high molecular weight chitosan, bleached fibers and activated carbon have been investigated. Zeolite recorded the highest adsorption capacity of 3.160 mg/g and reached equilibrium in two minutes. Carbon activation using KOH on tarap peels and rice husk may influence its porosity that contributed to higher adsorption capacity compared to bleached fibers.

Acknowledgments

Authors wishing to acknowledge Universiti Malaysia Sabah for funding this work (SDK 0044 – 2018).

References

[1] Hanum P, Ghani A, Yusoff K, Manaf L A and Bin Daud M 2009 Ammonium Ion Trend in Selected Malaysian River *World Appl. Sci. J.* **6** 442–8

[2] Dai Y, Dai Y, Sun Q, Wang W, Lu L, Liu M, Li J 2018 Utilizations of agricultural waste as adsorbent for the removal of contaminants: A review *Chemosphere* **11** 235–53

[3] Azreen I, Lija Y and Zahrim AY 2017 Ammonia nitrogen removal from aqueous solution by local agricultural wastes *IOP Conf. Ser. Mater. Sci. Eng.* **206** 12077

[4] Cechinel M A P, Ulson De Souza S M A G, and A. A. Ulson De Souza 2014 Study of lead (II) adsorption onto activated carbon originating from cow bone *J. Clean. Prod.* **65** 342–49

[5] Wang L 2012 Application of activated carbon derived from 'waste' bamboo culms for the adsorption of azo disperse dye: Kinetic, equilibrium and thermodynamic studies *J. Environ. Manage.* **102** 79–87

[6] Bernardi F, Zadinelo I , Alves H J, Meurer F and dos Santos L D 2018 Chitins and chitosans for the removal of total ammonia of aquaculture effluents *Aquaculture* **483** 203–12

[7] Zhou Y, Zhang L and Cheng Z 2015 Removal of organic pollutants from aqueous solution using agricultural wastes: A review *J. Mol. Liq.* **212** 739–62

[8] Binanipuram P and Kerala O 2016 Preparation and Characterization of Activated Carbon from Rice Husk **22** 551–8

[9] Wang X F, Zhang H P and Chen H Q 2006 *Preparation and characterization of high specific surface area activated carbon from bamboo by chemical activation with KOH* **37**

[10] Karadag D, Koc Y, Turan M and Armagan B 2006 Removal of ammonium ion from aqueous solution using natural Turkish clinoptilolite 2006 *J. Hazard. Mater.* **136** 604–9

[11] Huang H, Xiao X, Yan B and Yang L 2010 Ammonium removal from aqueous solutions by using natural Chinese (Chende) zeolite as adsorbent *J. Hazard. Mater.* **175** 247–52

[12] Delkash M, Ebrazi Bakhshayesh B and Kazemian H 2015 Using zeolitic adsorbents to cleanup special wastewater streams: A review *Microporous Mesoporous Mater.* **214** 224–41

[13] Beebe D A, Castle J W and Rodgers J H 2013 Treatment of ammonia in pilot-scale constructed wetland systems with clinoptilolite *J. Environ. Chem. Eng* **1** 1159–65

[14] Chung Y C, Li Y H and Chen CC 2005 Pollutant removal from aquaculture wastewater using the biopolymer chitosan at different molecular weights *J. Environ. Sci. Heal. - Part A Toxic/Hazardous Subst. Environ. Eng.* **40** 1775–90

[15] Korobochkin V V, Tu N V and Hieu N M 2016 Production of activated carbon from rice husk Vietnam *IOP Conf. Ser. Earth Environ. Sci.* **43** 012066

[16] Gaskin J W, Steiner C, Harris K, Das K C and Bibens B 2008 Effect of Low-Temperature Pyrolysis Conditions on Biochar for Agricultural Use *Trans. ASABE* **51** 2061- 9

[17] Satayeva A R, Howell C A, Korobeinyk, A V, Jandosov, J, Inglezakis, V J, Mansurov, Z A, Mikhalovsky S V 2018 Investigation of rice husk derived activated carbon for removal of nitrate contamination from water *Sci. Total Environ.* **630** 1237–45

[18] Zahrim A Y, Ricky L N S, Hilal N and Tamrin K F 2017 Ammonia-Nitrogen Recovery from Synthetic Solution using Agricultural Waste Fibers *Indian J. Sci. Technol.***10** 1–5

[19] Zahrim A Y, Ricky, L N S, Shahril Y, Rosalam S, Nurmin B, Harun A M and Azreen I 2015 Partly Decomposed Empty Fruit Bunch Fiber as a Potential Adsorbent for Ammonia-Nitrogen from Urban Drainage Water in *InCIEC 2014* (Singapore: Springer Singapore) p 989–1001.

[20] Rožić M, Cerjan-Stefanović Š, Kurajica S, Vančina V and Hodžić E 2000 Ammoniacal nitrogen removal from water by treatment with clays and zeolites *Water Res.* **34** 3675–81

[21] Zhou L and Boyd C E 2014 Total ammonia nitrogen removal from aqueous solutions by the natural zeolite, mordenite: A laboratory test and experimental study *Aquaculture* **432** 252–7

[22] Guo H, Zhang X Y and Liu J L 2016 Ion-exchange Capability for Ammonium Removal using Zeolite Modified by Potassium Permanganate *Chem. Eng. Trans.* **55** 63–168

IOP Conf. Series: Materials Science and Engineering **606** (2019) 012004 doi:10.1088/1757-899X/606/1/012004

Synthesis of nano emulsion from waste cooking oil for enhanced oil recovery applications

M I I M Noor[1], A Islam[3], S Saalah[1], C C Ken[1] , S M Anisuzzaman[1,2] and Z Kamin[1,2*]

[1]Chemical Engineering Programme, Faculty of Engineering, Universiti Malaysia Sabah, Jalan UMS, 88400, Kota Kinabalu, Sabah, Malaysia
[2]Energy Research Unit, Faculty of Engineering, Universiti Malaysia Sabah, Jalan UMS, 88400, Kota Kinabalu, Sabah, Malaysia
[3]Department of Petroleum and Mining Engineering, Faculty of Engineering and Technology, Jessore University of Science and Technology, Bangladesh
*Corresponding author: *zykamilia@ums.edu.my

Abstract. Enhanced oil recovery (EOR) is a tertiary method to extract potential oil in a matured well. Emulsion, which is one EOR method has properties which enable it to recover oil remains in reservoir rocks of the matured well. Nano sized emulsion or nano emulsion able to recover trapped oil, by capillary force in fine pores of the reservoir rock. Nano emulsion reduces the oil-water interfacial tension and therefore changes the wettability of the reservoir rock. However, the nano emulsion is uneconomical and unstable. Therefore, this work aimed to produce nano emulsions by using low cost waste cooking oil, Tween 80 and Span 80. Ultra-sonication was used to reduce the size of the emulsion. The concentration of surfactant, power of ultra-sonication and time of sonication were varied to study the stability of the oil-in-water and size of the emulsion using the Malvern Zetasizer. The nano emulsion was then tested on a crude oil soak meligan formation sand rock to test the performance of the nano emulsion. The result show values of zeta potential and droplet size were - 15.3 mV and 262.1 d.nm respectively. The emulsion droplet sizes were about 500 nm with milky white appearance. The nano emulsion shows positive results as 63.89 % crude oil displaced from the soaked meligan formation sand rock. As a conclusion, the nano emulsion produced from waste cooking oil can potentially be used for EOR.

1. Introduction

As the population increases, the demand of oil production is also increasing. The petroleum industry is facing difficult challenges on the demand of oil and it needs technological innovations to overcome it. Several techniques have been used to maintain a production of crude oil at maximum level. In an oil and gas field, once the primary recovery or natural pressure of a well decreases, human intervention is needed using secondary and followed by tertiary recovery. The EOR is a tertiary recover that can efficiently extract remaining oil trapped in the reservoir rock by capillary forces [1]. This phase uses fluid that consists of several components and miscible with oil, than can therefore reduce the viscosity and improve the flow of the oil. EOR also able to extract up to 60 – 65% of the remaining oil in a well [2] therefore increase production of oil as the demand increases.

In EOR, there are several techniques can be used to increase the extraction of oil depending on the reservoir. Each reservoir has its own characteristics. Nano technology is one of the unique solution that can be used to help and overcome this challenge due to its unique characteristics, efficient, and more environmentally sound technology. Nano emulsion also protects the layer of oil and gas to avoid permanent blockage in the oil and gas flow. Due to small droplet size, they are not subject to gravity-driven separation, but the density differences of the two phases. [3]. The nano emulsion that has been prepared is a homogeneous mixture of hydrocarbon and water with a ratio of surfactant [1].

Surfactant is a chemical substance that absorbs on or concentrated at a surface or fluid-fluid interface when present at low concentration in a system. The surfactant acts as an emulsifier to absorb the interface between water and oil medium by reducing the interfacial tension between them and prevent from conglomerate [4]. Nano emulsion is a transparent, translucent (50 – 200 nm) or milky white emulsion (about 500 nm) [5]. It is an efficient tool for EOR technique, because the emulsion has a high level of extraction efficiency by reducing oil-water interfacial tension down to <0.001 mN/m as resulted to solubilize the oil residual very efficiency which will increase the oil recovery and the completely clean up the formation [4]. However, the current nano emulsion is uneconomical and often unstable. Therefore, in this study, we proposed to produce emulsion in the size of nano scale. Focusing at formulation that can provide properties such as minimum average droplet size and minimum amount of surfactant, higher stability of nano emulsion and should be able to be used to displace oil from the pores of a rock.

2. Material and methods

2.1 Material

A filtered waste cooking oil had been obtained from a local food store. Distilled water and surfactants sorbitan monooleate (Span 80) and polyoxyethylenesorbitan monoleate (Tween 80) procured from Sigma-Aldrich Chemie Gmbh, Germany and from R & M Chemical, United Kingdom respectively were used without treatment. A meligan formation sand rock sample and crude oil were used to test the potential use of the nano emulsion in EOR.

2.2 Experimental procedure

2.2.1 Determination of concentration surfactant ratio

The combination of a high and low Hydrophilic-lipophilic balance (HLB) emulsifier is often more effective than the use of a single emulsifier. HLB is the balance of the size and the strength of the hydrophilic (water loving or polar) and the lipophilic (oil loving or non-polar) groups of the emulsifier. All emulsifier consist of a molecule that combine both hydrophilic and lipophilic groups. The value of HLB of an emulsifier is related to its solubility. Thus, an emulsifier having a low HLB will tend to be oil-soluble use for producing water-in-oil emulsions, and one having a high HLB will tend to be water-soluble (aqueous characteristics) is used for oil-in-water emulsions [6]. Figure 1 shows the HLB range.

HLB value	0	10	20
Classification	Oil-loving group	Balanced	Water-loving Group

Figure 1. HLB range.

Method for selection of surfactant as emulsifying agent is known as an HLB method by using the following equation to calculate the HLB mixed surfactant system:

$$HLB = \frac{m_A \times HLB_A + m_B \times HLB_B}{m_A + m_B} \tag{1}$$

Where m_A and m_B are the mass of surfactants A and B, respectively. HLB_A and HLB_B are the HLB number of surfactants A and B, respectively [6].

2.2.2 Value of surfactant ratio

A blend of a low HLB and a high HLB surfactant gives the HLB needed to form an emulsion and has better performance than a pure surfactant [6]. The HLB values for oil-in-water emulsion for this experiment, required 10 HLB for waste cooking oil (corn oil) [6]. By using, $HLB_{Span\ 80}$ and $HLB_{Tween\ 80}$ of 4.3 and 15.0 respectively in Equation 1 the % of Tween 80 and Span 80 obtained were 53.3 and 46.73% respectively. The ratio rounds off to 2.0 vol. Tween 80: 1.0 vol. Span 80. The most stable sample of emulsion can be formed with the formation of an emulsion system by using the mixture of Span-80 and Tween-80 surfactant in the ratio of 49/51 [7].

2.2.3 Preparation of nano emulsion

This experiment used high emulsification method to prepare nano emulsion. Distilled water and waste cooking oil were measured using a graduated cylinder and funnel to avoid spillage. The ratio of waste cooking oil to distilled water was set at one to four respectively [8]. As for the surfactant, pipette and measuring cylinder were used for small and higher amount of volume respectively. Volume for each material were shown in Table 1 which were filled in a 200 ml beaker. Oil and surfactant mixture was blended in a beaker and stirred mildly using a magnetic stirrer for 5 minutes. The required amount of water was simultaneously added slowly into the oil phase.

Table 1. Volume of emulsion raw material.

Exp.	Surfactant (ml)			Oil (ml)	Water (ml)	Total (ml)
	Span 80	Tween 80	Tot.			
1	1.67	3.33	5.0	30	120	155
2	3.33	6.67	10.0			160
3	5.00	10.00	15.0			165
4	6.67	13.33	20.0			170
5	8.33	16.67	25.0			175
6	10.00	20.00	30.0			180

2.2.4 Determination of ultrasonic intensity of the particle size of nano emulsion

The generated ultrasonic intensity form the Sonotrode horn tip was calculated using the equation 2, where, I is the intensity; P is the input power; and r, is the radius of the sonotrode horn tip [9]. The radius of Sonotrode horn tip is 2.54 cm. Based on the Equation 2, the corresponding ultrasonic intensities for power input of 150 to 750 W show in Table 2.

$$I = \frac{P}{\pi r^2} \tag{2}$$

Table 2. Conversion of power to ultrasonic intensity.

Input power (W)	Ultrasonic intensity (W/cm^2)
0	0
150	7.4007
300	14.8014
375	18.5018
450	22.2022
600	29.6029
750	37.0036

2.2.5 Ultra-sonication process to form droplet size of particle

Emulsion material was resized by using sonication process of Ultrasonic Processors (Model no. VC 750, Sonics & Materials, Inc.). The Sonotrode horn tip of ultrasonic was adjusted in the range of 1 to 0.5 cm depth from the bottom of the surface of the beaker. As a precaution, the timer was set a pulse to have a rest around 20 second for each 50 seconds of sonication because the instrument will heat up time to time. After sonication, the beaker needed to be cool down for 5 minutes before continuing with the Malvern Zetasizer instrument.

2.2.6 Measurement of droplet size
Malvern Zetasizer Nano ZS, was used to measure the particle size and zeta potential by Dynamic Light Scattering (DLS) of the nano emulsion. It is a measurement of the magnitude for electrostatic or charge repulsion/attraction between particles, which affect its stability. In general the particle with zeta potential more positive than +30mV or more negative than -30mV are normally considered stable. There are two types of cell that need to be used in the Malvern Zetasizer which were the disposable sizing cuvette and clear disposable zeta cell for measuring particle size (d.nm) and zeta potential (mV) respectively.

2.2.7 Determination of the performance of nano emulsion in EOR
The nano emulsion potential to be used for EOR will be determined by referring work by [5]. The nano emulsion was selected from the sample based on the smallest oil droplets and stable emulsion. A Maligan formation sand rock, which had been soaked in crude oil for a 10 day period, was immersed in the selected nano emulsion. Weight of nano emulsion and crude oil (g) and weight crude oil (g) were calculated using Equation 3 and 4 respectively

Weight of immersed Sand Rock (Crude Oil) in nano emulsion – Weight of sand rock after immersed in emulsion (4days) (3)

Weight of nano emulsion and crude oil – Weight of 200mL nano emulsion (4)

3. Result and discussion

3.1 Effect of surfactant concentration on particle size
To determine the effect of surfactant concentration on particle size, the amplitude and time of sonification were set at 100 % or 750 W and 15 minutes respectively. The obtained results were tabulated in Table 5. In general the droplet size decreases with the amount of surfactant. This is due to the increases in interfacial area and decreases in the interfacial tension [7].

Figure 2 shows that in general the higher amount of surfactant in the emulsion, resulted in the increase of particle size. For instance, at 14.29 vol. % of surfactant, the particle size shown the lowest value of 515.8 nm. Subsequently, the stability of the emulsion depends on the value of zeta potential. If the particles in the suspension have a large zeta potential regardless of the charges, the particles tend to repel each other and decrease the tendency for flocculattion to occur. In contrast, if the particles in the suspension have a low zeta potential values, then there is no force to prevent the particles to coming together and flocculating.

Figure 3 shows the highest value of zeta potential obtained was at surfactant concentration of 6.25 vol. %, making it the most stable emulsion compared to the others concentration at. In addition, physical observation of surfactant concentration of 6.25 vol. The % is shown in Figure 4 (b) in which, when compare with other emulsion, this emulsion does not seem to separate after a few days. Nevertheless, surfactant of 6.25 vol % was selected for the subsequent experiment instead of 14.29 vol. % to reduce costs by using the least amount of surfactant with a stable condition.

3.2 Effect of sonication time on particle size
The effect of sonication time on particle size was studied by fixing the amplitude of the ultrasonic at 100 % same as previous and surfactant concentration at 6.25 vol. %. The distilled water and waste cooking oil is in ratio 1 to 4 respectively.

Figure 5 shows that initially an increase in the time of sonication decreases the particle size. This indicates that it is possible to produce nano emulsion, using less energy and shorter time. At time 0, there was no sonication occurred, and after 5 minutes the particle size begins to decrease. This is

IOP Conf. Series: Materials Science and Engineering **606** (2019) 012004 doi:10.1088/1757-899X/606/1/012004

true for the time between 5 to 15 minutes. However, after 15 minutes as the sonication continues, the size of the particle increase significantly.

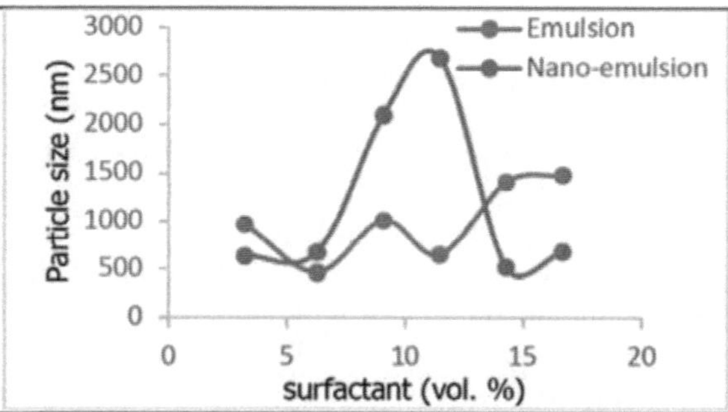

Figure 2. Graph particle size of the emulsion before and after sonication against surfactant concentration.

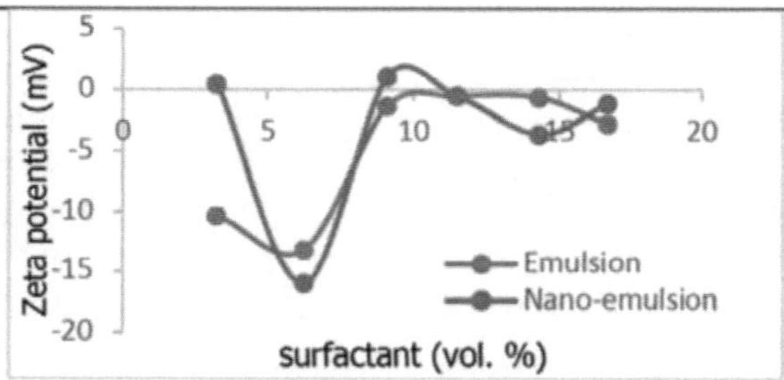

Figure 3. Graph zeta potential of the emulsion before and after sonication against surfactant concentration.

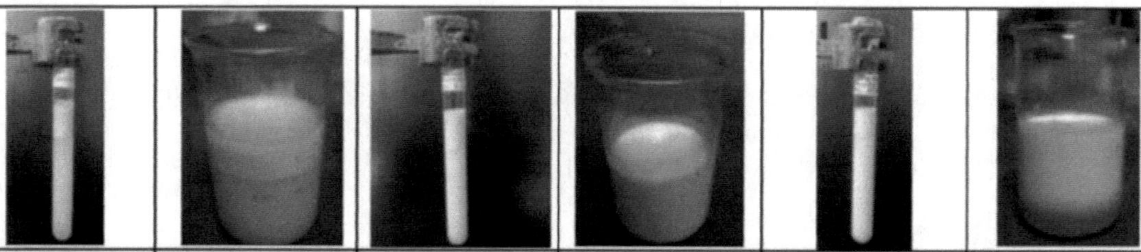

Emulsion	Nano emulsion	Emulsion	Nano emulsion	Emulsion	Nano emulsion
Particle size = 972.0 nm	Particle size = 632.3 nm	Particle size = 465.3 nm	Particle size = 672.5 nm	Particle size = 1010.3 nm	Particle size = 2100.0 nm
Zeta potential = -10.3 mV	Zeta potential = 0.49 mV	Zeta potential = -13.2 mV	Zeta potential = -16.00 mV	Zeta potential = -1.36 mV	Zeta potential = 1.09 mV
(a)		(b)		(c)	

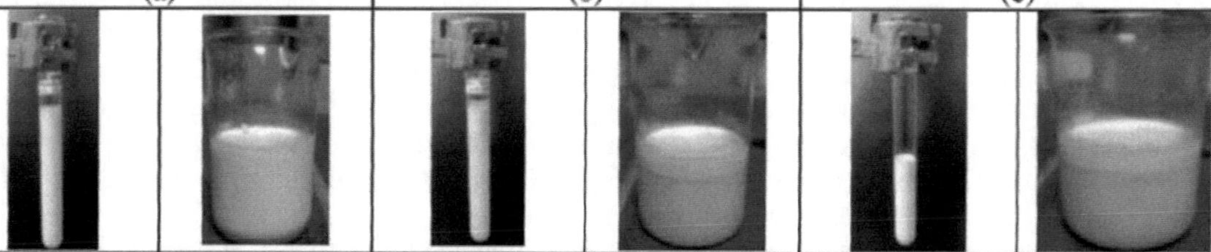

Emulsion	Nano emulsion	Emulsion	Nano emulsion	Emulsion	Nano emulsion
Particle size = 646.5 nm	Particle size = 2684.0 nm	Particle size = 1411.0 nm	Particle size = 515.8 nm	Particle size = 1485.0 nm	Particle size = 681.4 nm
Zeta potential = -0.54 mV	Zeta potential = -0.49 mV	Zeta potential = -0.63 mV	Zeta potential = -3.65 mV	Zeta potential = -2.80 mV	Zeta potential = -1.17 mV
(d)		(e)		(f)	

Figure 4. Concentration of surfactant (a) 3.23 vol.%; (b) 6.25 vol.%; (c) 9.09 vol.%; (d) 11.77 vol.%; (e)14.29 vol.%; and (f) 16.67 vol.%.

Figure 6 shows the zeta potential value. As the time of sonication increases, zeta potential also increases. An emulsion at stable state should have a zeta potential value of ± 30 mV. However the highest magnitude value obtained in this experiment was at - 0.485mV for 45 minutes sonication time. However, the corresponding value of particle size is bigger than the particle size of emulsion using five minutes sonication. Therefore the emulsion using the 5 minutes time of sonication was taken as the reference for the following experiment, due to a lower energy usage than the others and produces considerably small particle size. Additionally, Figure 7 (b) shown that there was no separation of the nano emulsion having 5 minutes time of sonication. The appearance shows no phase separation of oil and water, which means the emulsion is homogeneous.

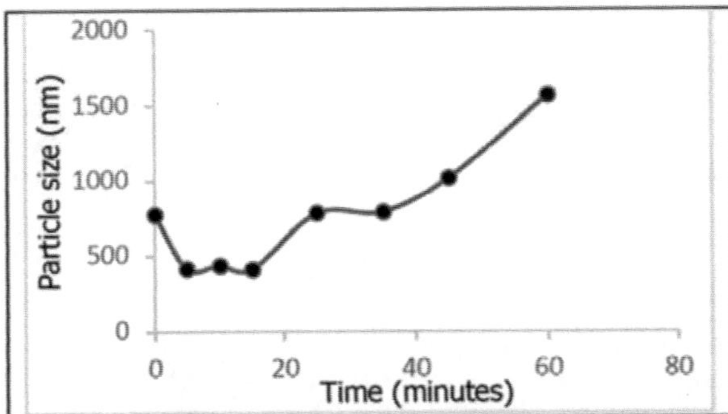

Figure 5. Graph size droplet particle of the nano emulsion against the time of sonication.

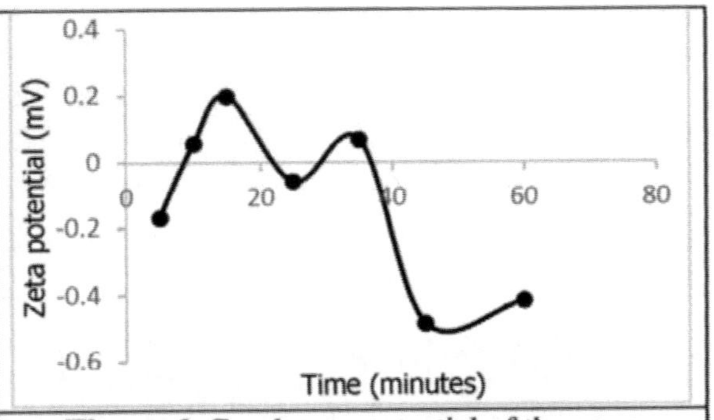

Figure 6. Graph zeta potential of the nano emulsion against the time of sonication.

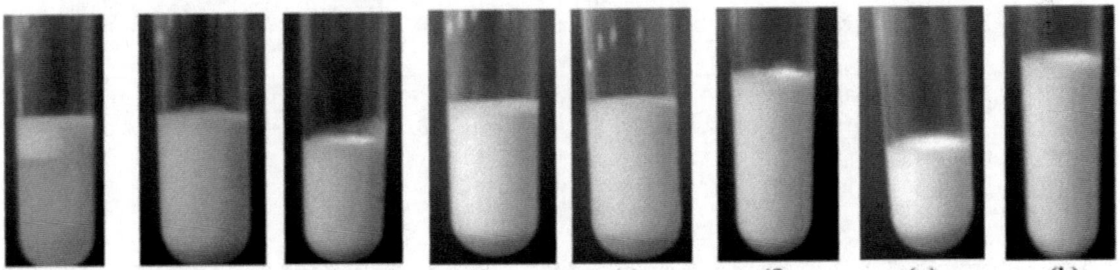

Figure 7. Time of sonication from 5 to 60 minutes (a) 0 minutes, (b) 5 minutes, (c) 10 minutes, (d) 15 minutes, (e) 25 minutes, (f) 35 minutes, (g) 45 minutes and (h) 60 minutes.

3.3 Effect of ultrasonic intensity on particle size

The experiment was done to investigate the role of sonication power in producing of smaller size of nano emulsion. The results obtained from previous experiment were used, which were 6.25 vol. % surfactant and 5 minutes time of sonication. The result of this experiment was plotted in Figure 8.

In general, at low power of 20 % the particle size is observed to reduce from 326 to 266 d.nm. However, after the 20 % power, the particle size was observed to increase with % of sonication power. This may due to the increase in shear stress in the liquid as the ultrasonic intensity increase. This shear stress can create a great number of cavitation of tiny bubbles in the liquid which then collapsed drastically to become small size droplet [9]. Subsequently, the small droplet then forms bubbles as the % of the sonication power increase.

The 20 % power of sonication shows a high value of zeta potential of - 15.3 mV nearing to a stable emulsion of ± 30 mV as shown in Figure 9. It is the highest value, compare to the other emulsions. Additionally, it uses less energy to produce the nano emulsion. However, after a few days, it does not show stability as shown in Figure 10 (a) when compared to other nano emulsion in Figure 10 (b-f) where a distinct separation of phases of the nano emulsion can be observed.

Figure 11 shows that, the highest peak of median bell graph of the size of the particle for the nano emulsion of 20% power sonication. The % intensity vs. particle size distribution curves has been shown for clarity.

IOP Conf. Series: Materials Science and Engineering **606** (2019) 012004 doi:10.1088/1757-899X/606/1/012004

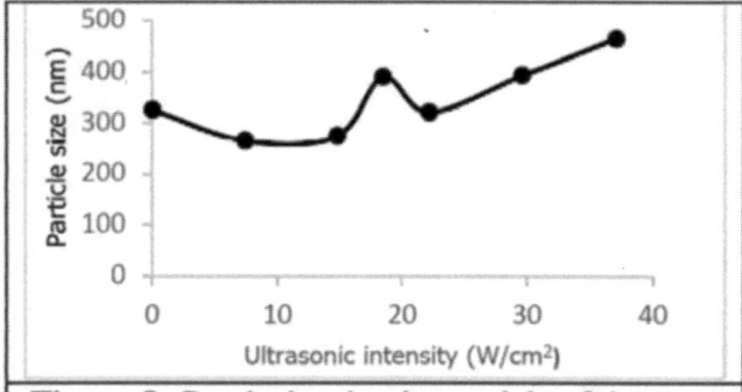

Figure 8. Graph size droplet particle of the nano emulsion against the ultrasonic intensity.

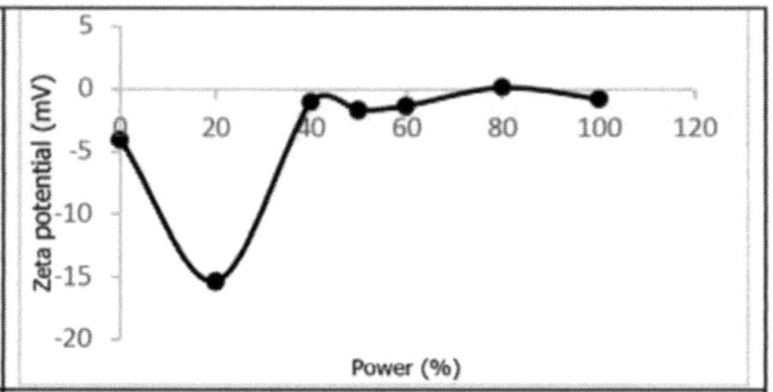

Figure 9. Graph zeta potential of the nano emulsion against the power of sonication.

Figure 10. Power of sonication to form nano particle size at power (a) 20 %, (b) 40 %, (c) 50 %, (d) 60%, (e) 80 and (f) 100 %.

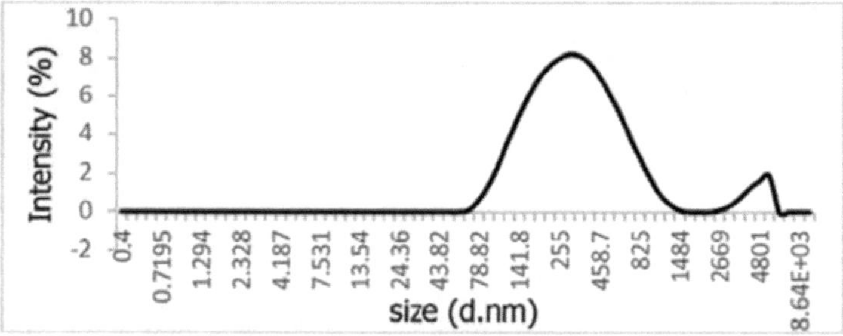

Figure 11. Particle size distribution for 20% power sonication.

3.4 Use of a nano emulsion for EOR application

The nano emulsion of 20 % power sonication was selected for further investigation for EOR application. Initial investigation shows that the mixture of the nano emulsion and crude oil (Figure 12 (a) was separated after several minutes, with crude oil on top of the mixture as seen in Figure 12 (b).

Meligan formation sand rock at an initial weight of 63.4 g were soaked with crude oil for 10 days [5]. The weight of the sand rock after the immersion was 67 g in which the rock had absorbed about 3.6 g of crude oil after 4 days. This indicates that the rock may potentially have high porosity.

Subsequently, the crude oil soaked Meligan formation sand rock was later immersed in the selected nano emulsion weigh at 184.1g (200 ml). The results were recorded in 1 hour, 2 and 4 days as shown in Figure 13. By using Equations 3 and 4, the results show about 2.3 g crude oil appear on the surface of nano emulsion after being immersed for 4 days. In which, about 63.89 % of oil was being pulled out by the nano emulsion from the sand rock.

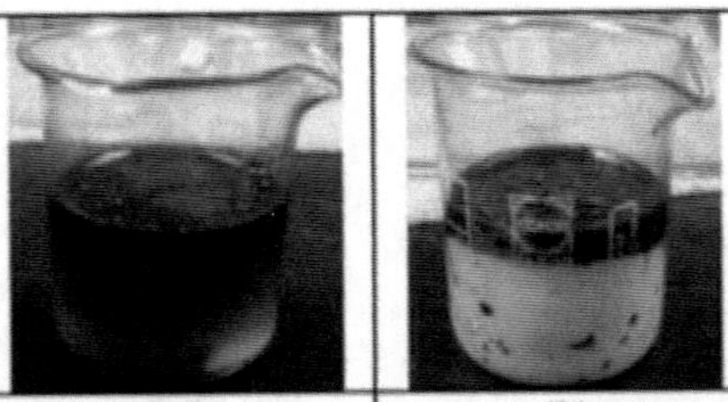

(a) (b)

Figure 12. (a) Mixture of nano emulsion and crude oil, (b) Result after several minutes.

(a) (b) (c)

Figure 13. Meligan formation sand rock immersed in a nano emulsion (a) 1 hour, (b) 2 days and (c) 4 days.

Nano emulsions having a droplet size, which is smaller than the pores of reservoir rock shows potential to displace the crude oil trapped inside the pore of reservoir rock. The unique characteristics of nano emulsion that has been formulated with surfactant and waste cooking oil, which has smaller particle size, enable the nano emulsion to reduce the interfacial tension and provide more surface contact with the crude oil that lead to the increase of the efficiency of the displacement process [10].

4. Conclusion and recommendation

In this study, a series of nano emulsion had been prepared by varying the surfactant concentrations, time and power of ultra-sonication. The result shows that, by using 6.25 vol. % surfactant and 20 % sonication power and 5 minutes of sonication time can result in the formation of a nano emulsion with reduced droplet size of 262.1 d.nm and zeta potential of -15.3 mV. The EOR application test showed that the nano emulsion able to extrate crude oil at about 63.89 %. It is an indication that the nano emulsion can be potentially used to recover residual oil, which is still trapped inside the fine pore of reservoir rocks. Further research can be carried out to investigate the type of surfactant or type of oil that can affect the nano emulsions particle size and stability since the result of this research, does not achieve a stable emulsion as indicated by zeta potential values. A study on the synthesis of nano emulsion needs consideration in every aspect, to achieve a lower cost, efficiency, stability and environmental friendly for EOR.

References

[1] Bera A and Mandal A 2015 Microemulsions: a novel approach to enhanced oil recovery: a review. *Journal of Petroleum Exploration and Production Technology*

[2] Tunio S Q, Tunio A H, Ghirano N A, Mohamed Z and Adawy E 2011 Comparison of Different Enhanced Oil Recovery Techniques for Better Oil Productivity. *International Journal of Applied Science and Technology*. **1** (5), 143–153

[3] Mandal A, Bera A, Ojha K and Kumar T 2012 Characterization of Surfactant Stabilized Nanoemulsion and Its Use in Enhanced Oil Recovery. in: *SPE Int. Oilf. Nanotechnol. Conf. Exhib*

[4] Hashimah Alias N, Aimi Ghazali N, Amran Tengku Mohd T, Adieb Idris S, Yahya E and Mohd Yusof N 2015 Nanoemulsion Applications in Enhanced Oil Recovery and Wellbore Cleaning: An Overview. *Applied Mechanics and Materials*

[5] Xu Q, Li Q, Wu X, Li B, Yao E, Li Y, et al. 2017 Enhancement of the Wettability and Lubrication of Shale Rock via Nanoemulsions. *International Journal of Polymer Science*

[6] ICI Americas Inc. (1976) The HLB System. *Chemmunique*. 37 (10), 2–22

[7] Porras M, Solans C, González C, Martínez A, Guinart A and Gutiérrez J M 2004 Studies of formation of W/O nano-emulsions. in: *Colloids Surfaces A Physicochem. Eng. Asp*

[8] Shrestha A 2011 Effect of span 80 - tween 80 mixture compositions on the stability of sunflower oil-based emulsions

[9] Shahavi M H, Hosseini M, Jahanshahi M, Meyer R L, and Darzi G N 2015 Evaluation of critical parameters for preparation of stable clove oil nanoemulsion. *Arabian Journal of*

IOP Conf. Series: Materials Science and Engineering **606** (2019) 012004 doi:10.1088/1757-899X/606/1/012004

Chemistry
[10] Effah Y, Nur H A, Nurul A G, Tengku A T M and Hafizzuddin O 2015 Nanoemulsion Formulation Using Biodegradable Oil in Enhance Oil Recovery (EOR) *Applied Mechanics and Materials*. **754–755** 1098–1101

IOP Conf. Series: Materials Science and Engineering **606** (2019) 012005 doi:10.1088/1757-899X/606/1/012005

Bond Strength of Corroded High Yield Steel Bars Embedded in Normal Strength Concrete

T Thomas, A K Mirasa and N Rizalman
Faculty of Engineering, University Malaysia Sabah, Kota Kinabalu, Malaysia

E-mail: fasttrexx@gmail.com

Abstract Experimental investigation on the relationships between corrosion of reinforcement and concrete for the bond strength in pullout test and flexural test has been conducted to established the maximum strength. The reinforcing bars used in this study were rusted for four, eight and twelve months, it is cleaned for the eight and twelve months before embedded in concrete cubes for pullout test and concrete beams for the flexural test. The first component for experimental to make reinforce bars exposed to weathers and the parameter included 20 N/mm^2 concrete compressive strength. The diameters of reinforcing bars 12 *mm*, 16 *mm* and 25 *mm* used for tensile test and pullout test. 8 *mm*, 12 *mm*, 16 *mm* and 25 *mm* for flexural test. Tensile test carried out according to ASTM A615 16 and MS146:2006. Pullout test conducted according to ACI and BS 1881 Part 207:1992. Flexural test performed according to ACI and BS 8110 Part 1:1985, BS 1881 Part 109:1983 and BS 1881 Part 118:1983. All specimen's workmanship for reinforcement carried out according to BS 8110 Part 1: 1985 and to support this thesis based on previous literature review.

1. Introduction

This study is part of a long-term research program started in 2016 in Sandakan, Sabah and it is focused on corrosion without the concrete and corroded bars embedded in concrete for bond strength. Experimental tests on the tensile test were conducted in order to study corrosion level and the effects of natural corrosion on high yield steel bars. Pullout test and flexural test were conducted in order to study corrosion level that contributes to bond strength and bond stress strength. The initial specimen for 600 *mm* and 800 *mm* long, zero month consisted of three number none corroded high yield steel bars 12 *mm ø*, 16 *mm ø* and 25 *mm* and four, eight and twelve months specimens consisted of three number corroded high yield steel bars 12 *mm ø*, 16 *mm ø* and 25 *mm ø*. Maximum bond strength for tensile test and pullout test are taken for an average. Initial specimens 1400 *mm* long, zero month consisted of none corrode high yield steel bars 12 *mm ø*, 16 *mm ø* and 25 *mm ø* ties with 8 *mm ø* stirrups and four, eight and twelve months specimens consisted of corroded high yield steel bars 12 *mm ø*, 16 *mm ø* and 25 *mm ø* ties with 8 *mm ø* stirrups. Maximum bond stress strength for the flexural test taken independently. All the results took based on zero, four, eight and twelve month's analyses.

2. Objectives and Significance of the Study

The objectives of this study to investigate the effects of corrosion rates on corroded high yield steel bars for tensile strength. To investigate the bond strength between corroded high yield steel bars and concrete on pullout test and flexural test. The significance of this study, to obtained knowledge regarding tensile, pullout and flexural test for its strength.

IOP Conf. Series: Materials Science and Engineering **606** (2019) 012005 doi:10.1088/1757-899X/606/1/012005

3. Methodology

12 *mm ø*, 16 *mm ø* and 25 *mm ø* high yield steel bars, and 8 *mm ø* mild steel bars stirrups from newly exposed to weathers for four, eight and twelve months until it is corroded. All of the specimens keep above ground and under the shed to control the corrosion level. The specimens are shown in (Figure 1, Figure 2 and Figure 3).

Figure 1. 600 *mm* long steel For tensile test.

Figure 2. 800 *mm* long steel Bars for pullout test.

Figure 3. 1400 *mm* long steel Bars with stirrups for the flexural test.

3.1. Corrosion Mechanism without Concrete

Steel bar are most common and useful materials use for reinforcing concrete but it has had one major flaw that it is corrodes as it is made almost entirely iron. Corrosion occurs when the iron (*Fe*) surface undergoes simple changes. First is the ($Fe \rightarrow Fen+ + n\ electrons$) iron atom loses some electrons and become positively charged, an ion allows it to bond with another group of atoms that is negatively charged. Second, wet iron rust to give a variant of iron oxide so the other parts of reactions involved of oxygen (O_2) and water (H_2O) or ($O_2 + 2H_2O + 4e- \rightarrow 4OH^-$). Negative (-) charged the materials, that which combined with the electron and the specimens (high yield steel bars) which are produced in the first reaction [1] show in Equation 1 and Equation 2.

$$2Fe + O_2 + 2H_2O \rightarrow 2Fe(OH)_2 \tag{1}$$

Iron (High Yield Steel Bars) + Oxygen + Water → Iron Hydroxide dissolved in it. Oxygen dissolves in Water and reacts with Iron Hydroxide caused Corrode.

$$4Fe(OH)_2 + O_2 \rightarrow 2H_2O + 2Fe_2O_3H_2O \tag{2}$$

Iron Hydroxide + Oxygen → Water + Hydrated Iron Oxide (Brown colour corrode).

3.2. Tensile Test

The tensile property is an integral part of describing the mechanical properties of materials and it is applied to every engineering disciplines. Tensile properties found from tensile test include stress, strain, elastic modulus and the ultimate tensile strengths of the material specimen. The others properties determined from the tensile test are yield strength, elongation percentage at a fracture and the percentage reductions in the area at its fractures [2].

$$\sigma = \frac{F}{A_o} \tag{3}$$

Stress (σ) determined by average loads or force (F) applied to sample divided by the original cross-sectional area (Ao) of the specimen. When the material subjected to uniaxial forces the specimen extends in direction of force, result displacement defines;

$$\varepsilon = \frac{\Delta L}{L_o} \tag{4}$$

strain ($\mathcal{E}$) or the ratio between the changes in length (ΔL) of the sample in direction of force over the original length (L_o) of the sample

$$\Delta L = L - L_o \tag{5}$$

L = final length of the material specimen. The stress-strain relationship elastic modulus

$$E = \frac{\sigma}{\varepsilon} \tag{6}$$

σ = Stress
$\mathcal{E}$ = Strain
The ultimate tensile strength,

$$\sigma u = \frac{Fmax}{A_o} \tag{7}$$

The highest load achieved at any point throughout the test (F_{max}) it must be divided by the original cross-sectional area (A_o) [3].

3.3. *Pullout Test*

Bond stress is calculated as average stress between the reinforcing bar and the surrounding concrete along the embedded length of the bar. In general, bond stress corresponding to maximum pull out load that is can be regarded as bond strength or the ultimate bond. The criterion of ultimate bond strength is characterized by its clear definition and simplicity in bond strength interpretation [4]. Uniform bond, bond stress S can be expressed;

$$S = \frac{Pmax}{(L \times d)} \tag{8}$$

$Pmax$ = maximum pull out load
L = Embedded bar length
d = diameter of the bar
The equation was employed in the present calculation of bonding stress between the embedded steel bar and the surrounding concrete for the specimen.

3.4. *Flexural Test*

The flexural test is a method for evaluating the strength of a brittle material due to break of the sample around specimen gripping [5]. The test employs in the rectangular beam and the specimen bent until fracture or yield using a three-point flexural test technique. The flexural strength represents the highest stress at its moment of yield, it measures in term of stress given symbol "σ."

$$\sigma = \frac{3FL}{2bd^2} \tag{9}$$

F = Load (force) at the fracture point
L = Length of the support span
b = width
d = thickness

IOP Conf. Series: Materials Science and Engineering **606** (2019) 012005 doi:10.1088/1757-899X/606/1/012005

4. Finding

Specimens of 12 *mm ø*, 16 *mm ø* and 25 *mm ø* high yield steel bars productions from factory show that all specimens weight and area are variant. It is revealed from the observation after exposed to weathers, rust with no serious flaking begins at four months and reduced in weight and area at the eight months and twelve months and it shows pitting corrosion on the surface. Due to none uniform of natures, corrosion distribution of tensile behaviors on corroded specimens for tensile results obtained from tensile UTM 1000, universal testing machine, show that all specimen strength results for tensile is more than expected and it shows the strength is above 600 N/mm^2 which is stronger than [6]. This also shows that the strength of a material depends on the quality productions from factory and corrosion level or condition of materials is dependent on the weathers. Hence, it is significant none corrode materials to be used to reinforce the concrete, data shown in Table 1.

Table 1. Weight, Area, Maximum Stress Strength and Percentage Reduction in Weight and Area of 12 *mm ø*, 16 *mm ø* and 25 *mm ø*, 600 *mm* Long High Yield Steel Bars for Tensile Test Results.

No.	Designation of Specimens	Weight (gram)	Reduced		Area (mm^2)	Reduced		Maximum Stress Strength (N/mm^2)
			(gram)	(%)		(mm^2)	(%)	
1	T12-0	526.67	-	-	111.82	-	-	628.5
2	T12-4	523.33	-	-	111.11	-	-	618.0
3	T12-8	520.67	0.67	0.13	110.55	0.15	0.13	617.4
4	T12-12	530.67	0.67	0.13	112.67	0.14	0.12	622.5
5	T16-0	920.00	-	-	195.33	-	-	657.4
6	T16-4	926.70	-	-	196.74	-	-	652.4
7	T16-8	910.67	0.67	0.07	193.35	0.14	0.07	646.2
8	T16-12	913.67	0.34	0.04	196.99	0.07	0.04	646.7
9	T25-0	2293.33	-	-	486.91	-	-	638.6
10	T25-4	2283.33	-	-	484.78	-	-	618.2
11	T25-8	2263.67	0.34	0.02	480.61	0.07	0.02	663.5
12	T25-12	2300.67	0.67	0.03	488.50	0.18	0.03	665.5

Results show after high yield steel bars is cleaned it reduced in weight and area on the eight and twelve months. 12 *mm* diameter specimens P12-8 and P12-12 both reduced by 0.14% in weight and 0.13% in area, bond strength is 22 N/mm^2 promotes bond strength than P12-0 control specimen for 20 N/mm^2. 16 *mm* diameter specimens P16-8 and P16-12 both reduced by 0.05% in weight and 0.06% in area, bond strength is 19 N/mm^2 has the same bond strength to P16-0 control specimen. 25 *mm* diameter specimens P25-8 and P25-12 both reduced by 0.02% in weight and 0.02% in area, bond strength is 13 N/mm^2 has the same bond strength to P25-0 control specimen. Steel bars used in this study relatively not effective on the bond strength for pullout test on the four months of corrosion as bond strength did not promote for P12-4, P16-4 and P25-4 as bond strength are 17 N/mm^2, 12 N/mm^2 and 8 N/mm^2 shown in Table 2.

Table 2. Weight, Area, Maximum Bond Strength and Percentage Reduction in Weight and Area of 12 *mm ø*, 16 *mm ø* and 25 *mm ø*, 800 *mm* Long High Yield Steel Bars for Pullout Test.

No.	**Designation of Specimens**	**Weight** (gram)	**Reduced** (gram)	(%)	**Area** (mm^2)	**Reduced** (mm^2)	(%)	**Maximum Bond Strength** (N/mm^2)
1	P12-0	711.67	-	-	113.33	-	-	20
2	P12-4	711.33	-	-	113.27	-	-	17
3	P12-8	711.67	1	0.14	113.33	0.15	0.13	22
4	P12-12	711.67	1	0.14	113.33	0.15	0.13	22
5	P16-0	1251.33	-	-	199.25	-	-	19
6	P16-4	1251.67	-	-	199.31	-	-	12
7	P16-8	1251.67	0.67	0.05	199.31	0.11	0.06	19
8	P16-12	1251.67	0.67	0.05	199.31	0.11	0.06	19
9	P25-0	3099	-	-	493.47	-	-	13
10	P25-4	3099	-	-	493.47	-	-	8
11	P25-8	3098.67	0.67	0.02	493.42	0.11	0.02	13
12	P25-12	3098.67	0.67	0.02	493.42	0.11	0.02	13

Results show after high yield steel bars is cleaned it reduced in weight and area on the eight and twelve months. 8 *mm* diameter stirrups for specimens F8-12-8, F8-16-8, F8-25-8 reduced by 0.34% in weight and 0.34% in area and specimens F8-12-12, F8-16-12 and F8-25-12 reduced by 0.4% in weight and 0.4% in area. 12 mm diameter specimens F12-8 and F12-12 both reduced by 0.08% in weight and 0.08% in area, bond strength is 19 N/mm^2 has the same bond strength to the P12-0 control specimen. 16 *mm* diameter specimens F16-8 and F16-12 both reduced by 0.05% in weight and 0.04% in area, bond strength is 18 N/mm^2 has the same bond strength to the FB16-0 control specimen. 25 *mm* diameter specimens F25-8 and F25 -12 both reduced by 0.01% in weight and 0.01% in area, bond strength is 27 N/mm^2 has the same bond strength to the F25-0 control specimen. Steel bars used in this study relatively not effective on the bond strength for flexural test on the four months of corrosion as bond strength did not promote for F12-4, F16-4 and F25-4 as bond strength are 11 N/mm^2, 13 N/mm^2 and 17 N/mm^2 shown in Table 4.

Table 3. Weight, Area and Percentage of Reduction in Weight and Area for 8 *mm ø* 100 *mm* X 150 *mm* X 100 *mm* X 150 *mm* Mild Steel Stirrups for Reinforced Concrete Beams 1500 *mm* X 150 *mm* X 200 *mm*.

No.	Designation of Specimens	Weight (gram)	Reduced (gram)	Reduced (%)	Area (mm^2)	Reduced (mm^2)	Reduced (%)
1	F8-12-0	197	-	-	50.19	-	-
2	F8-12-4	197	-	-	50.19	-	-
3	F8-12-8	197	0.67	0.34	50.19	0.17	0.34
4	F8-12-12	197	0.78	0.4	50.19	0.2	0.4
5	FB8-16-0	197	-	-	50.19	-	-
6	F8-16-4	197	-	-	50.19	-	-
7	F8-16-8	197	0.67	0.34	50.19	0.17	0.34
8	F8-16-12	197	0.78	0.4	50.19	0.2	0.4
9	F8-25-0	197	-	-	50.19	-	-
10	F8-25-4	197	-	-	50.19	-	-
11	F8-25-8	197	0.67	0.34	50.19	0.17	0.34
12	F8-25-12	197	0.78	0.4	50.19	0.2	0.4

Table 4. Weight, Area, Maximum Bond Stress Strength and Percentage Reduction in Weight and Area of 12 *mm ø*, 16 *mm ø* and 25 *mm ø*, High Yield Steel Bars ties with 8 *mm ø* 100 *mm* X 150 *mm* X 100 *mm* X 150 *mm* Mild Steel Stirrups Reinforced Concrete Beams 1500 *mm* X 150 *mm* X 200 *mm* for Flexural Test.

No.	Designation of Specimens	Weight (gram)	Reduced (gram)	Reduced (%)	Area (mm^2)	Reduced (mm^2)	Reduced (%)	Maximum Bond Strength (N/mm^2)
1	F12-0	1267	-	-	115.29	-	-	19
2	F12-4	1267	-	-	115.29	-	-	11
3	F12-8	1267	1	0.08	115.29	0.09	0.08	19
4	F12-12	1267	1	0.08	115.29	0.09	0.08	19
5	F16-0	2219	-	-	201.91	-	-	18
6	F16-4	2219	-	-	201.91	-	-	13
7	F16-8	2219	1	0.05	201.91	0.09	0.04	18
8	F16-12	2219	1	0.05	201.91	0.09	0.04	18
9	F25-0	5453	-	-	496.18	-	-	27
10	F25-4	5453.75	-	-	496.25	-	-	17
11	F25-8	5454	0.75	0.01	496.27	0.07	0.01	27
12	F25-12	5453.5	0.75	0.01	496.23	0.07	0.01	27

[7, 8] points out light rust is not usually a problem with the reinforcements and as it creates a roughened surface that may promote bonding. Figure 6 shown rust with no serious flaking on high

yield steel bars surface, embedded in concrete. As it assumed light rust with light loose scale. The result showed a weaker bond strength on the fourth month. As the presence of corrode, it impairs the bond strength of the high yield steel bar because corrosion occurred and fills at raised ribs and gaps between the ribs, thus it is out from the original shape of high yield steel bar. The reduction of the mechanical lock by corrosion results in high yield steel bar raised ribs and gaps between ribs, decline the bond strength with concrete. In essence, the bond strength originates from the mechanical lock are between the raised ribs and concrete. The presence of corrosion on raised ribs and gaps between ribs reduces the effective cross-sectional area of high yield steel bar. Hence, the available bond strength capacity of high yield steel bars and concrete is reduced by a considerable reduction in the bond strength shown in Figure 4 and Figure 5.

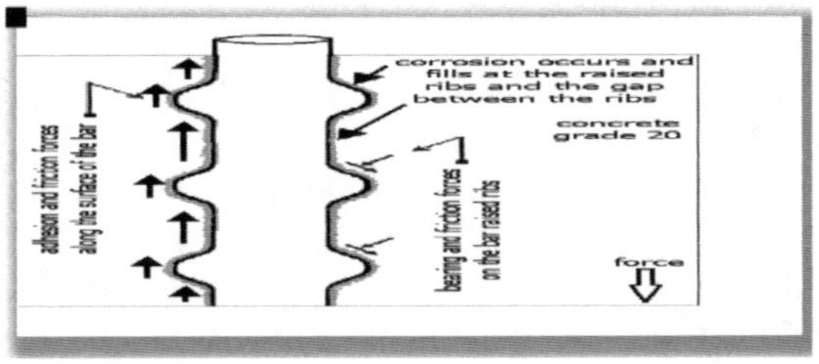

Figure 4. Bond-Stress generated on a high yield steel bar.

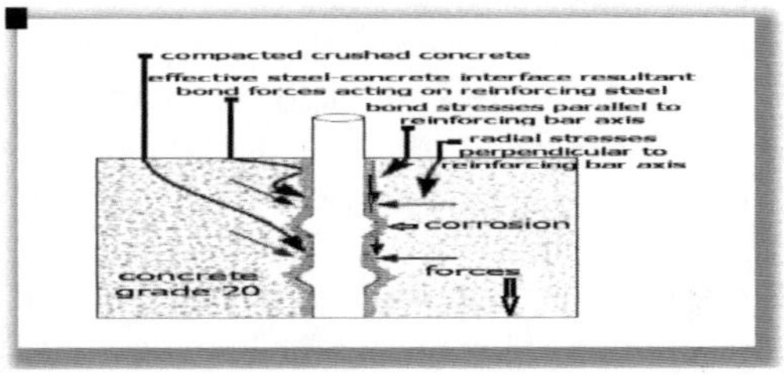

Figure 5. Force parallel and perpendicular high yield steel bar concrete interface steel bar.

[7, 8] points out all bars or wires shall be well and cleanly rolled and shall be sound and free from the surface defects and pipe, or other defects detrimental to it subsequent processing and to it end use. Rust, seams, surface irregularities or mill scale shall not the causes for rejection provided, as hardwire brushed test specimen fulfils all the requirement of the specifications. Figure 8 shows a corrode is clean before high yield steel bars embedded in concrete, as corrode with losing scale is clean, it showed pitting holes on high yield steel bars surface and reduced in weight and area. The result showed a promoted bond strength on eight and twelve months, as concrete grip high yield steel bars ribs and with the contributions from pitting holes on the high yield steel bars surface. The maximum bond strength data for pullout test and maximum bond stress strength data for flexural test shown in Figure 7 and Figure 9.

Figure 6. Light corrosion.

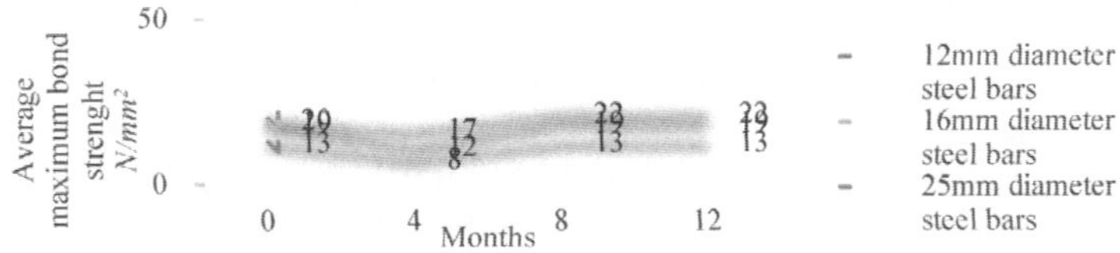

Figure 7. Maximum bond strength for the pullout test.

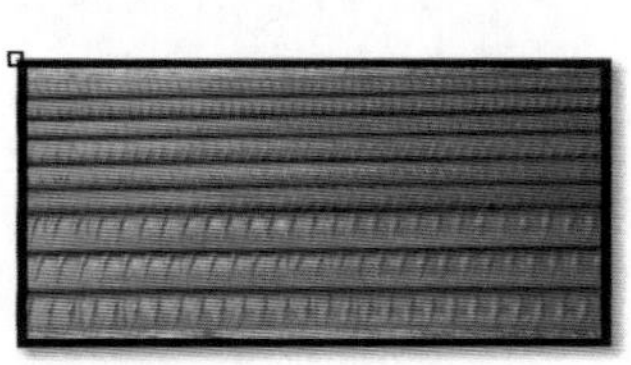

Figure 8. Steel bars clean reduced in weight and area, pit holes on the surface.

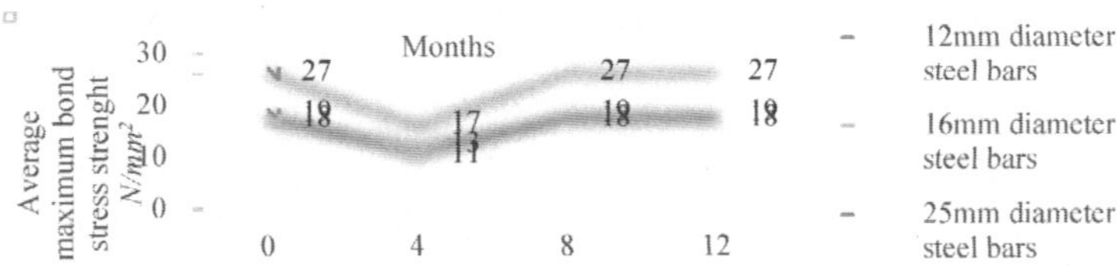

Figure 9. Maximum bond stress strength for flexural test.

5. Conclusion

The effect of corrosion on the high yield steel bars for tensile strength, pullout bond strength and flexural bond stress strength significant findings from this study, found and shown that the following conclusions can be drawn from the present study that all results taken based on zero, four, eight and twelve months analyses comprised corrosion level subjected exposed to the weathers and concrete strength showed the results that bond strength was not influenced by corrosion on the eight and twelve months exposure. The evidence from this study and the results of this study indicates that these findings as shown the effect of corrosion on the high yield steel bars for tensile strength, pullout bond strength and flexural stress bond strength from this analysis can be drawn:

a. Corrosion of high yield steel bars did not reduce the tensile strength of high yield steel bars for the periods of four, eight and twelve months.
b. Corrosion (light rust) of high yield steel bars impairs the bond strength between high yield steel bars and concrete at the periods of four months.
c. Corrosion of high yield steel bars pairs the bond strength between high yield steel bars and the concrete at the periods of eight and twelve months.
d. Corrosion of high yield steel bars did not reduce the capacity of high yield steel bars in tensile strength.
e. Corrosion of high yield steel bars did not reduce the capacity of high yield steel bars in the reinforced concrete due to pullout bond strength and flexural bond stress strength.

Acknowledgement

The author would like to thanks to Professor Ir. Dr Abdul Karim Mirasa, Dr Nurfaidhi Rizalman, Dean of the University Malaysia Sabah and Datuk Ir. Hj. Amrullah Hj. Kamal Director of Public Works Department Sabah and Vyner Sikodol Chief Assistant Director of Public Works Department Sabah for their assistance in completed my study in their laboratory gratefully appreciated.

References

[1] Nihal A T and M M 2015 Study of the behaviour of corroded steel bar and convenient method of repairing *HBRC J1* **12(2)**: 107-113.
[2] Craig R R, 1996 *Mechanics of Materials* John Wiley & Sons: New York, NY
[3] Dowling N E 1993 Mechanical Behaviour of Materials *Engineering Method for Deformation, Fracture and Fatigue, 4th Edn.* Marcia J.H., Pearson Education Limited 2013, Edinburgh Gate, Harlow, Essex CM202JE, England: 190-225
[4] Hadi M N S 2008 Bond of high strength concrete with high strength reinforcing steel *Open Civil J1* **2**:143-147.
[5] Ashby M F and Jones D R H 1980 *Engineering Materials: An introduction to their properties and application* 1st Edn. Pergamon Press.
[6] *Malaysian Standard (MS) 146: 2006.* Hot rolled steel bars reinforced concrete specification specified the characteristic strength is 460 N/mm^2:1-4.
[7] *ACI 318-2002. Building code requirement for structural concrete, American Concrete Institute* ACI 318 -02 Committee and Commentary 318R-02 USA.
[8] *ACI 318-2005. Building code requirement for structural concrete and commentary* ACI 1995 Farmington Hills MI. ACI Str. J1 89(6): 692-698 ASTM G 1, Standard practice for preparing, cleaning and evaluating corrosion test specimens: 15-21

IOP Conf. Series: Materials Science and Engineering **606** (2019) 012006 doi:10.1088/1757-899X/606/1/012006

Carbon dioxide capture using waste tire based adsorbent

K Duduku[1], S M Anisuzzaman, Bono, B K Teo and V V B Rao[2]
[1]Chemical Engineering Program, Faculty of Engineering, Universiti Malaysia Sabah, 88400 Kota Kinabalu, Sabah, Malaysia
[2]College of Technology, Osmania University, Hyderabad-500007-India

E-mail: krishna@ums.edu.my

Abstract. In this study, a waste tire derived activated carbon (AC) was assessed as a solid sorbent for CO_2 capture. The objectives of this study is to determine the effectiveness of AC from waste rubber tire via chemically activation with potassium hydroxide (KOH) as a low cost solid adsorbent on the adsorption of CO_2 from the mixture of N_2 and CO_2 at different adsorption temperature, adsorption flow rate and impregnation ratio. The reactor temperatures were 30°C, 40 °C and 50°C and the flow rate was 50 cm^3/min, with an impregnation ratio of 1 and 1.5. The breakthrough time was found to decrease with increasing of impregnation ratio of KOH and decreasing of adsorption temperature and flow rate. The AC with 1.5 KOH impregnation ratio has shown the best CO_2 adsorption capacity among the adsorbents studied and the adsorption capacity is 238.7mg/g of adsorbent at 30°C.

1. Introduction

Increasing carbon dioxide (CO_2) emission in environment leads to global warming which is an issue of great concern today. The climate of the earth is varying continuously due to various factors, for example: change in the earth's orbit, change in the Sun's intensity, change in ocean currents, volcanic emissions and increase in greenhouse gas (GHG) concentrations. The greenhouse effect is the phenomenon where water vapour, CO_2, methane and other atmospheric gases absorb the outgoing infrared radiation causing an increase of earth's temperature. Excessive greenhouse gases in the atmosphere are responsible for various environmental problems like continuous rise of water-level in sea, increasing number of ocean storms, floods, and others [1]. Among the greenhouse gases, CO_2 contributes more than 60% to global warming because of its huge emission amount [2].

Carbon capture and sequestration (CCS) is widely acknowledged and identified to be the major and crucial approach to limit the CO_2 level in the realm of carbon reduction and its derived technologies. CCS encompasses the removal and sequestration of CO_2 from industrial low CO_2 level containing streams, the generation of high concentrated CO_2 stream, and the transport and storage of compressed CO_2 [3,4]. Currently, three types of CCS processes for scrubbing exist: pre-combustion, oxyfuel combustion and post-combustion. Post combustion capture technology has a higher thermal efficiency than pre combustion.

Various materials such as zeolites and activated carbons (AC) have been investigated as adsorbents for the CO_2 separation and capture by selective adsorption. Among these materials, AC is the most commonly used and most effectively modified because of its high specific surface area and big pore volume[5-7]. However, the cost of AC is relatively high, which limits its application. Researchers are now finding other adsorbents as an alternative to the standard activated carbons, which have lower cost and similar efficiency. Natural materials or the waste /by products of industries or synthetically prepared materials , which cost less and can be used as such or after some minor treatment as adsorbents are generally called low-cost adsorbents. Environmental issues due to solid waste management are common problems arise in developing countries. One of the increasing wastes in the dumping site is stockpile of waste rubber tires. Li et al. [8] stated that the waste tires can contribute to health and environmental problems. The accumulation of waste rubber tires required large spaces, thus results in large spaces in landfill. In addition, due to the stockpile of tires, it can be a breeding place for mosquitos when stagnant water was collected insides the tire and eventually it could cause a spreading of diseases including dengue fever and encephalitis. One of the ways to reduce the stockpile of waste tires is by using it as raw material to prepare activated carbon for CO_2 adsorption properties. AC prepared from rubber tire has been used widely in the adsorption process [9].

Thus, In this work a cost effective, low energy intensity, and high capacity adsorbent from activated carbon derived from waste tire for CO_2 adsorption was developed. The carbon dioxide was adsorbed from a mixture of CO_2/N_2 since flue gas emissions are often associated with considerable amount of nitrogen gas. AC from waste rubber tire via chemical activation with potassium hydroxide (KOH) is a new substance to be tested its suitability as CO_2 adsorbent. The impregnation ratio of potassium hydroxide (KOH) is used for modification of waste tire derived AC. The adsorption experiment involved study of the effect of some parameters such as effect of adsorption temperature and adsorption flow rates on the breakthrough curves.

2. Experimental Methods

2.1 Preparation of activated carbon

The raw material used in this experiment was a waste rubber tire, which was cut into small particle sizes ranging around 3-5mm, and was washed with distilled water, then dried in an oven for 24 hours at 80°C. 100g of the waste rubber was weighted using an analytical balance. The sample was heated in the high temperature muffle furnace at temperature of 500°C for 2 hours. 20g char obtained was impregnated with potassium hydroxide. The impregnated potassium hydroxide to char was divided into 1.0 and 1.5 ratios. The samples were then placed in a conical flask. 150 ml of distilled water was added and the flasks were agitated using an orbital shaker for 24 hours at constant speed. Thereafter, the KOH impregnated samples were dried in an oven at 110°C for 24 hours. Then, 10g impregnated samples were heated again using the high temperature muffle furnace at temperature 600 °C for 3 hours. After the burning process was completed, the activated carbon obtained was washed using 1M hydrochloric acid (HCl) to remove ash and dried in an oven at 110°C for 24 hours.

2.2. Burn off and yield percentage

The burn off and yield percentages were calculated as:

$$\text{Burn off} = \frac{Y-X}{Y} \times 100\% \tag{1}$$

$$\text{Product yield} = \frac{X}{Y} \times 100\% \tag{2}$$

where X = weight of activated carbon and Y = weight of rubber tire to prepare AC.

2.3. Determination of ash content

Approximately 1.0 g of sample weighed in a beaker and the dried sample was transferred into a tarred crucible. The sample ignited until it was completely burned using muffle furnace at 500°C for 4 h and then it was cooled in desiccators. The loss in weight is obtained and the ash content is calculated by using the following equation :

IOP Conf. Series: Materials Science and Engineering **606** (2019) 012006 doi:10.1088/1757-899X/606/1/012006

$$\% \, of \, ash \, content = \frac{Weight \, of \, sample \, after \, ash \, process \, x \, 100\%}{Weight \, of \, sample \, before \, ash \, process} \tag{3}$$

2.4. Determination of moisture content

According to Standardization and Industrial Research Institute Malaysia (SIRIM) method, 1.0 g of sample weighed in a tarred weighing beaker to the nearest 10 mg and placed in a drying oven for 2 h at 110°C. The moisture content was determined as :

$$\% \, of \, moisture \, content = \frac{Loss \, of \, weight \, x \, 100\%}{Weight \, of \, sample \, after \, drying} \tag{4}$$

2.5. Determination of pH

Approximately 1.0 g of AC was put inside a 250 mL Erlenmeyer flask. Then, 100 mL of distilled water was poured inside the Erlenmeyer flask. The solution was heated for 5 min in gentle boiling condition. Later, the solution was cooled to room temperature and diluted with distilled water to 100 mL. Solution was stirred well and the pH was determined using pH meter (Mettler Toledo MC 226, Switzerland) .

2.6. Particle size distribution

The particle size distribution of samples was determined by Malvern Zen 3600 analyzer which isbased on laser diffraction technique in the operation. For the analysis of particle size, 0.1g of sample was dissolved in 20 ml distilled water in 2 hours before conducting the analysis. After that, a little of the solution was pipetted into a clear and clean cuvette and put into the analyzer.

2.7. Pore Diameter and Surface Area

Automated Mercury Porosimeter, Pascal 440 series was used to determine the pore size and surface area of sample. It involved the intrusion of mercury at high pressure into a material. The pore size can be determined based on the external pressure needed to force the liquid into a pore against the opposing force of the liquid's surface tension.

2.8. Estimation of amount of CO_2 adsorbed

The amount of CO_2 adsorbed was calculated from the breakthrough curves using the following equation.

$$t = \int_0^{\alpha} \left(1 - \frac{c}{c_0}\right) dt \tag{5}$$

2.9. CO_2 adsorption experiment

The adsorption experiments were carried out in our laboratory designed a jacketed fixed bed flow reactor. Schematic diagram of experimental setup is shown in Figure 1. For each experiment about 2.00 g AC was loaded into the reactor. Simulated flue gas mixture of N_2 and CO_2 was used to study of CO_2 adsorption. Initial concentration was measured by CO_2 meter. N_2 and CO_2 gases from two different cylinder stocks were passed separately through mass flow controllers at total flow rate of 50 cm^3/min with 1 % of CO_2 and were mixed together before passing through reactor. Then the gas mixture (simulated flue gas) and water from the water bath were passed through the adsorbent bed at different inlets. The reactor temperature was precisely controlled by using water bath controllers. For each one minute, the outlet concentration of CO_2 was measured by passing the reactor effluent through a CO_2 meter. The water was recycled back to the water bath. The CO_2 capture experiment was stopped immediately once the CO_2 concentration was achieved their saturated concentration. The CO_2 concentration value was obtained from the CO_2 meter. The experiment was repeated with different impregnation ratios of AC, temperatures and flow rates. Parameters for CO_2 adsorption experiment are presented in Table 1.

Table 1. Parameters for CO_2 adsorption experiment

Sample	Gas flow rate (cm^3/min)	Temperature (°C)	Pressure (atm)
AC 6 -1.0		30	
AC 6-1.5	50	40	1
		50	
AC 6 -1.0		30	
AC 6-1.5	80	40	1
		50	
AC 6 -1.0		30	
AC 6-1.5	100	40	1
		50	

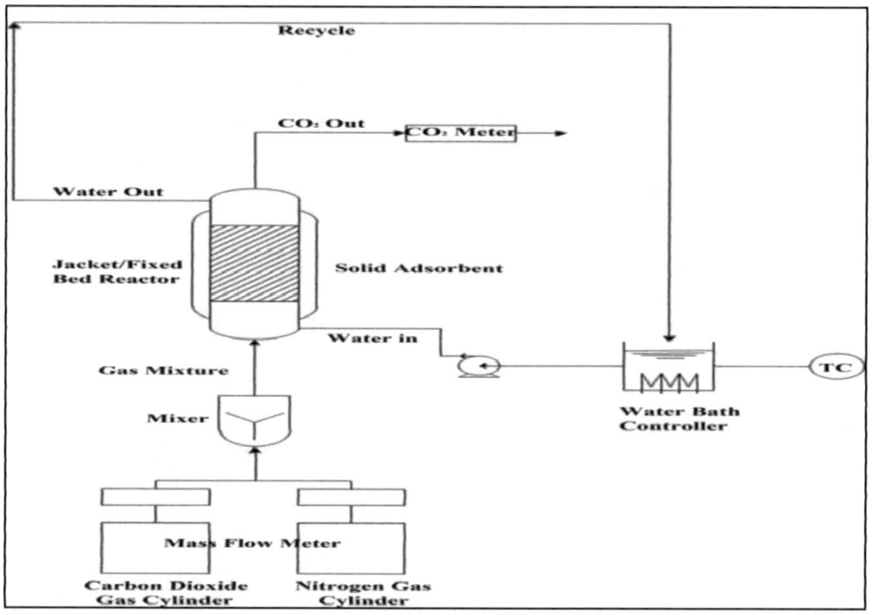

Figure 1. Schematic of experiment setup

3. Results and discussion

3.1. Characterization of waste tire derived AC

Results from proximate and ultimate analyses of the waste tire derived AC are shown in Table 2. From the Table 2, the highest observable percentage yield is 96.7% at 600°C with 1.0 impregnation ratio of KOH. The effect of activation temperature and chemical impregnation ratio on yield was studied as all the variables were found to be significant terms for AC yield. It was concluded that the carbon yield decreased with increasing activation temperature. Thus, 600 °C was preferred than 800 °C for the activation. The increase in temperature would release increasing volatiles. As a result, intensifying dehydration and elimination reaction will occur. Moreover, it quickened the gasification reactions of carbon and therefore, attacks the amorphous components which obstruct the pores and causes a decrease in the carbon yield .

Table 2. Ultimate proximate analysis of waste tire derived AC

Sample name	AC 6-1.0	AC 6-1.5
Activation temperature °C	600	600
Impregnation ratio	1.0	1.5
Burn off (%)	3.28	11.09
Product yield (%)	96.7	88.9
Moisture content (%)	5	4
Ash content (%)	3	2
pH	6.02	6.86
Particle diameter, nm	170.56	170.50

The percentage of burn off for the sample of AC 6-1.0 and AC 6-1.5 are 3.28% and 11.09% respectively (Table 2). The AC yield was also strongly affected by the chemical impregnation ratio where increasing impregnation ratio decreased the yield and increased the carbon burn-off. This was because when higher impregnation ratio was used, the weight losses were due to increasing release of volatile products as a result of intensifying dehydration and elimination reactions also increased.

Furthermore, moisture content is the amount of water on the activated carbon under normal condition. The practical limit for the level of moisture content allowed in the activated carbon varies within 3 to 6%. The moisture content for the AC 6-1.0 and AC 6-1.5 are 5 % and 4% respectively. Moisture in the sample is very important since it significantly reduces the amount of CO_2 adsorbed. The presence of water vapour hinders CO_2 adsorption on activated carbon. The ash is present in the raw waste tire. Lower the ash content, the better is the activated carbon. The practical limit for the level of ash content allowed in the activated carbon which varies from 2 to 5%. From the Table 2, the percentage of ash content for AC 6-1.0 and AC 6-1.5 were 3% and 2% respectively, which are within the standard limit. The particle diameter for AC 6-1.5 was 170.5nm which is 0.06 smaller than AC 6-1.0. The early breakthrough observed with larger particle sizes of AC 6-1.0 was attributed to smaller mass transfer coefficient due to smaller specific surface area which also revealed the surface rated dependency of the sorption system. Empirically, it can be said that adsorption performance of CO_2 on the waste tire derived activated carbon was improved as the adsorbent particle sizes were reduced.

3.2. Effect of impregnation ratio

Figure 2 shows the breakthrough curve at **30**°C with different impregnation ratios of KOH of the waste tire derived AC. The activated carbon was impregnated with KOH at the ratio of 1 and 1.5. Dinda [10] proposed that CO_2 uptake capacity increases with the increase of impregnation ratio of activating agent. To know the effect of KOH impregnation ratio on CO_2 adsorption capacities, the experiment was carried out at 2 different impregnation ratios ranging from 1 to 1.5 at 30 °C with a flow rate of 50 cm^3/min. From the Figure, it can be clearly observed that the AC with impregnation ratio 1.5 showed a greater CO_2 adsorption since the higher impregnation ratio increases the effective surface area and actives sites for CO_2 adsorption. According to Tan et al.[11], at high KOH impregnation ratio, the pore development was mostly due to the intercalation of potassium metal in the carbon structure. As the impregnation ratio of KOH increased, the catalytic oxidation also caused the widening of pores.

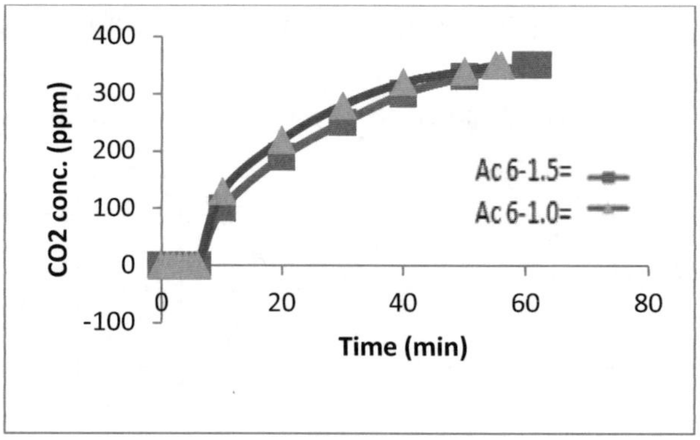

Figure 2. Breakthrough Curve at **30**°C with different impregnation ratio of KOH

3.3. Effect of temperature on adsorption

Figure 3 (a,b), shows the CO_2 breakthrough curve at 50 cm^3/min, for 2 different samples.

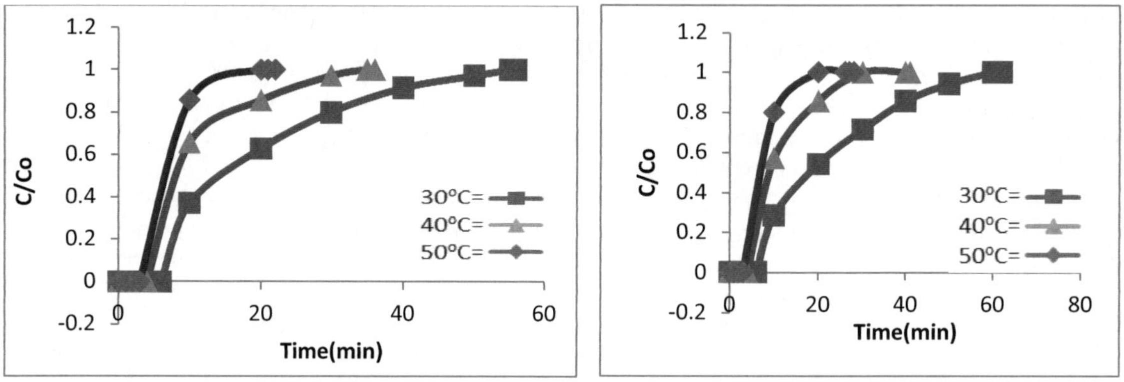

(a) (b)

Figure 3. CO_2 breakthrough curve at 50 cm^3/ min (a) AC^-1.0 (b) AC6 -1.5

Based on CO_2 capture result, it can be observed that when the temperature is increased, the CO_2 and nitrogen breakthrough times are shorter due the exothermic character of adsorption. Thus, the gas is very fast in achieving saturated concentration on their surface as temperature increases. Based on the Figure 3, for the sample AC 6-1.0, the shortest breakthrough time is 22 min by at 50 cm^3/min gas flow rate.

3.4. Effect of flow rate

Figure 4 shows the CO_2 capture with AC 6-1.5 at different gas flow rates. Based on the CO_2 capture result, increasing of gas flow rate can decrease the breakthrough time of CO_2 gas to achieve the saturated concentration. At the flow rate of 50 cm^3/min, the breakthrough time was 62 min elongated as compared with the higher flow rate of 100 mL/min which was 38 min at the temperature of 30°C. This result was supported by the previous researcher which said that, as the gas inlet flow rate increased, the breakthrough time also will decrease and causes degree of CO_2 adsorption to decrease. In Figure 6, only AC 6-1.5 was selected as adsorbent to undergo adsorption process as it has shown better performance in terms of adsorption capacity, more available active sites and has low particle size which contributes to longer breakthrough time.

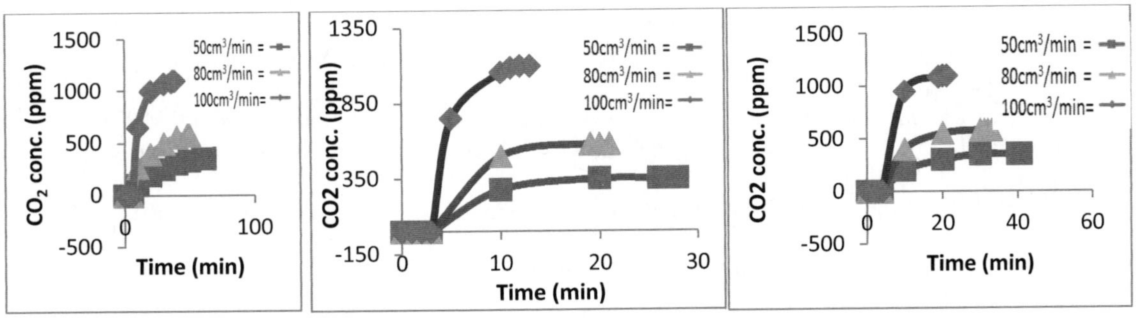

(a) (b) (c)

Figure 4. CO_2 capture of AC6-1.5 with different gas flow rates(a) at 30^{0}C (b) 40^{0}C (C) at 50^{0}C

As shown in Table 3, the highest removal percentage can be found in AC 6-1.5 which is 52.9% at temperature of 30 °C and initial concentration of CO_2 350 ppm. Higher impregnation ratio with KOH introduced some molecules of potassium on its surface which enhanced the adsorption of CO_2. This has proven that at lower temperature, the efficiency in capturing CO_2 is higher. Furthermore, as temperature increases from 30 °C to 50 °C, the removal percentage decreases.

Table 3. Removal percentage of CO_2

Sample name	Initial concentration of CO_2, C_o (ppm)	Removal percentage (%)		
		30 °C	40 °C	50 °C
AC6-1.0	350	50.0	21.4	2.86
AC6-1.5		52.9	25.7	5.71
AC6-1.0	580	39.7	13.8	0.9
AC6-1.5		46.6	19.0	12.1
AC6-1.0	1100	31.8	11.8	0.9
AC6-1.5		35.5	23.6	1.8

4. Conclusion

Adsorptive capture of CO_2 from flue gases is a promising technology for greenhouse gas mitigation. Low cost solid sorbent with high surface area is expected to offer the benefits for the chemical activation of waste tire derived activated carbon with KOH mostly used for CO_2 absorption process. From the experimental results, the breakthrough time decreases as temperature and flow rate increases. It can be concluded that an adsorbent with 1.5 KOH impregnation ratios has the highest CO_2 adsorption capacity amongst the studied adsorbents. AC 6-1.5 has removal percentage of 52.9 % and it has CO_2 adsorption capacity of 238.7 mg/g of adsorbent at 30°C and 50cm^3/min .

Acknowledgement
The authors would like to express greatest appreciation to the Centre of Research and Innovation (PPI) for the financial support.

IOP Conf. Series: Materials Science and Engineering **606** (2019) 012006 doi:10.1088/1757-899X/606/1/012006

References

[1] Lu J, Cheng M, Ji Y and Zhang H 2009 *J. Fuel Chem. Technol.* **37** 740-746
[2] Albo A, Luis P and Irabin A 2010 *Ind. Eng. Chem. Res.* **49** 11045–11051
[3] Krishnaiah D, Bono A, Anisuzzaman S M, Joseph C, Khee T B and Aaron D 2005 *J. Appl. Sci.* **14** 321-348
[4] Zhao A 2012 *University of Alberta Libraries* Alberta(USA) https://doi.org/10.7939/R3T07Z
[5] Krishnaiah D, Anisuzzaman S M, Joseph C G, Daud, W and Sundang M 2017 *Korean J. Chem. Eng.* **34** 1377-1384
[6] Krishnaiah D, Bono A, Anisuzzaman S M, Joseph C and Khee, T B 2014 *J. Appl. Sci.* **14** 3142-3148
[7] Joseph C G, Bono A, Krishnaiah D, Ling C Y, Ban N C, Anuar K and Joseph C G 2001 *J. Appl. Sci.* **9** 21-24
[8] Li L, Liu S and Zhu T 2010 *J. Environ. Sci.* **22** 1273-1280
[9] Edward L, Danny C and Gordon M 2004 *Carbon* **42** 2789-2805
[10] Dinda S 2013 *Sep. Purif. Technol.* **109** 64-71
[11] Tan I W, Ahmad A L and Hameed B H 2008 *Desalination* **225** 13-28

IOP Conf. Series: Materials Science and Engineering **606** (2019) 012007 doi:10.1088/1757-899X/606/1/012007

Nutrients Recovery Processes for Sewage: A Short Review

A Y Zahrim[1*], N N Safie[1], N Bolong[1], M Rajin[1], N M Ismail[1], S Saalah[1], M H M Yusop[2], C M Ongkodun[3], L N Y Ricky[4]

[1] Faculty of Engineering, Universiti Malaysia Sabah, Jalan UMS, 88400 Kota Kinabalu
[2] Faculty of Food Science and Nutrition, Universiti Malaysia Sabah, Jalan UMS, 88400
Kota Kinabalu
[3] Biotechnology Research Institute, Universiti Malaysia Sabah, Jalan UMS, 88400 Kota
Kinabalu
[4] Genting SDC Sdn. Bhd., Wisma Genting Plantations, KM12, Jalan Labuk, 90000 Sandakan, Sabah.

E-mail: zahrim@ums.edu.my

Abstract. The abundance of untreated sewage that consists of valuable nutrients and water scarcity has motivated the community to process and recycle the waste through efficient and sustainable approach. The presence of dissolved nutrients and emerging contaminants are increasing lately due to the rapid anthropogenic activities. Over the years, many studies on sewage processing have been reported in the literature: bioaugmentation, membrane bioreactor, moving bed biofilm as well as adsorption technique for nutrients recovery. The advantages and limitations of these technologies are discussed in brief. Finally, the treated sewage that still contain useful nutrients can be recycled as irrigation water for crops and non-crops plantation, toilet flushing etc. Nutrients that were adsorbed in the filter can be utilized by plant when it is used as growing media.

Keywords: sewage processing, recycling, irrigation, process water

1. Introduction

In Malaysia, despite of existing legislation that protects water quality, there are variety of pollutants emerge increasingly and require serious attention. Based on the Malaysia Environmental Quality Report [1], 43 % of rivers in Malaysia which included Sabah and Sarawak area are slightly polluted and 50 % rivers are recorded to be polluted due to the high biological oxygen demand (BOD), suspended solid (SS) and ammonia nitrogen (NH_3-N) which mainly contributed by improper sewage discharge. Furthermore, several studies conducted by Ashraf et al. [2], rare earth elements were detected in core marine sediments of coastal East Malaysia such as cesium at (16.56 - 36.84) mg/kg and heavy metals. Pharmaceuticals residue are also detected along the coastal area in Selangor which is originated from the sewage discharge [3]. Hence, a better and proper sewage treatment is necessary to reduce the waste and environmental pollution as well as to protect public health.

Sewage treatment remains challenging in terms of removing the unwanted dissolved contaminant particles that are small in diameter or size [4]. Consequently, several advanced sewage processes have been explored in this paper: bioaugmentation, membrane bioreactor (MBR), moving bed biofilm reactor (MBBR) and adsorption.

2. Sewage characteristics

Sewage (domestic wastewater (DWW) or municipal wastewater (MWW)) contains mainly human excreta or wastewater from human activities mainly from toilet, restaurants, washing activities, kitchen water and etc. Sewage contains microbial pathogens that are excreted in the stools or washing water at high concentrations [5]. On top of that, sewage also contains important carrier medium for nutrients in the nutrient cycle such as nitrogen and phosphorus [6].

Sewage is characterized as low-strength wastewater with chemical-oxygen demand (COD), 250-800 mg/L; SS, 120 – 400 mg/L [7]. Characteristics of sewage in various countries are summarized in Table 1. The range of BOD, COD, ammonium – nitrogen (N-NH_4^+), total dissolved solid (TDS), total suspended solids (TSS) are 123 – 979.62 mg/L, 212- 7303 mg/L, 22 – 135.2 mg/L, 115 – 439 mg/L, 75.8 - 623 mg/L respectively.

Table 1. Characteristics of sewage in various countries

Country	BOD	COD	TP	TDS	TN	Turbidity (NTU)	$N-NH_4^+$	TSS	NO_3^-	pH	*E.coli* (CFU /100ml)	Ref.
Selangor, Malaysia	NA	212	NA	115	NA	60	NA	76	NA	7.2	NA	[8]
Johor, Malaysia	980	7303	NA	NA	NA	919	NA	NA	NA	5.2	NA	[9]
Nagaoka City, Japan	240	532	NA	NA	NA	NA	22	262	NA	NA	NA	[10]
Harare, Zimbabwe	527	297	18	NA	NA	NA	24	219	NA	7.0	NA	[11]
Sao Carlos City , Brazil	NA	436	14	NA	41	NA	NA	176	NA	6.9	NA	[12]
Cracow , Poland	200	361	NA	NA	NA	NA	135	NA	NA	8.3	1.4×10^8	[13]
Vashi, India	123	227	NA	NA	NA	NA	NA	167	NA	6.7	NA	[14]
Palestine	641	1267	13	439	NA	NA	58	623	NA	7.4	3.9×10^7	[15]

BOD, COD, TP, TN, $N-NH_4^+$, NO_3^- are in mg/L

3. Sewage processing systems

3.1 Bioaugmentation

Bioaugmentation can be defined as adding endogenous or indigenous microorganisms, either selected strains/mixed cultures to wastewater to improve the catabolism of specific compound [16]. Bioaugmentation can be implemented in two schemes; (1) in situ augmentation, which provides internal process enhancements that add catabolism capability in the aerobic compartment by addition of immobilized selected strain of microorganism; and (2) external bioaugmentation, which augments the sludge with external selected strain of microorganism grown in a separate reactor [17]. Microorganisms that have been used in previous studies are microalgae [18], commercial bacteria [19], photosynthetic bacteria [20] and oleaginous bacteria [21].

Bioaugmentation is not limited to only aerobic digestion but also anaerobic digestion. However, implementation of any bioaugmentation strategy must be preceded by a successful isolation and identification of individual strains or microbial consortia that possess the unique metabolic capabilities required to biodegrade or biotransform target micropollutants[22]. Also, the disadvantageous effects were largely reported being associated with bioaugmentation failures caused by the lack of acclimation for microorganisms to survive under harsh environmental conditions [16].

3.2 *Moving-bed bioreactor (MBBR)*

Aerobic MBBR is equipped with the biofilm carrier movement produced by the agitation. Anoxic and anaerobic biofilm reactors are useful for slow growth rate organisms like nitrifiers [23] [24]. These reactor have been proven to be capable for carbon oxidation, nitrification, denitrification and deammonification [25][26]. Furthermore, separated process such as denitrification and nitrification can be carried out simultaneously in MBBR that will reduce space, cost and retention time[24]. MBBR can also be installed inside the sewer pipeline to enhance the removal of organic pollutants[27]. Type of media has been used such as polyurethane sponges [28], biodegradable polymer (PCL) [29], polyethylene (Aqwise ABC5 carrier), high-density polyethylene ring, Kaldnes (K1), porous polyurethane cuboid sprayed with activated carbon (BIOCONS)[30] with ammonia removal > 90 %.

3.3 *Membrane bioreactor (MBR)*

MBR integrates membrane filtration into the activated sludge process for sewage. The use of the filters which can be submerged inside the bioreactor or arranged in dedicated external vessels has extensive effects on the process and the design of wastewater treatment which has two core benefits such as high effluent quality and low footprint [31]. MBR has already been utilized as a useful option to remove soluble organics, nutrients, and micropollutants from wastewater [32]. However, the membrane permeability has reported to decrease during successive filtration cycle due to the inorganic components that will seriously blocked membrane pores and physical and chemical cleaning, which then increasing the operating costs due to downtime incurred[33][34].

Efforts have been made to optimize the membrane process, such as development of cheaper membrane material, improvement of membrane life circle[35] and improvement of hydraulic condition in membrane tanks etc. [36][37], [38]. The modification of pores filtration membranes used in MBR is up to ultra and microfiltration membranes with a pore size of 0.2 μm or less which not only retain bacteria but also viruses[39]. Also, by installing an anoxic zone in MBR has impressive impact in reducing COD by 95 % and nitrification was complete and up to 82% [39]. Instead of that, magnetic activated sludge (MAS) has also been used to help mitigate the membrane fouling and it is inert and biocompatible that will not be detrimental to biological activity[40]. Recently, anaerobic membrane reactors (AnMBRs) for MWW has attracted attention. [41].

3.4 *Adsorption*

Adsorption has been used to treat volatile organic compounds (VOC), ester [42], heavy metal ions [43], synthetic dyes[44]. Adsorption is more suitable for wastewater treatment as compared to conventional methods that include less space requirement particularly in comparison to biological treatment, flexibility in design and operation, lower sensitivity to process variation, greater efficiency, insensitivity to toxic pollutants, no formation of harmful substances and wider applicability because it can remove or minimize different types of pollutants [45].

According to Meng et al., (2018), adsorption mechanism can be divided into three which are physisorption, chemisorption and ion-exchange adsorption. The principal types of adsorbents include activated carbon, bioadsorbent, synthetic polymeric, and silica-based adsorbents [47]. Nutrient removal from wastewater has been experimented by Kim et al. [48], where their experiment used an amine-grafted adsorbent for simultaneous phosphate ions (PO_4^{3-}) and nitrate ion (NO_3^-) ,and found out that the synthesized absorbent recovered 78% nitrate ion and 93% phosphate ion from wastewater.

Fruit waste adsorbent has been used by Zahrim et al. [49] to treat NH_3-N in synthetic wastewater which follows pseudo-second order with adsorption capacity, Q_e of 1.2 mg/g. Low cost adsorbent such as langsat peels, jackfruit seeds, and moringa peels has been used for ammonia-nitrogen adsorption. The highest adsorption capacity was recorded by langsat peels with 3.58 mg/g followed by jackfruit seeds and moringa peels with 3.37 mg/g and 2.64 mg/g respectively[50]. The submerged adsorbent can be used as media for plant growth later (Figure 1).

Figure 1. Lemon grass (*Cymbopogon citratus*) plant growth using different ratio of soil-submerged adsorbent: a) Bag No. 1: 100 % of soil, b) Bag No. 2: 50 % of soil + 50% of submerged adsorbent, c) Bag No. 3: 75% of soil + 25 % of submerged adsorbent.

Although a large number of studies have been performed to figure out the efficiency in adsorption by removing organic and inorganic pollutants there has not been any single report that provides a full comparative analysis between different adsorption processes. As a result, the adsorption process can be referred to be effective but the application is subjected only to remove a specific type of pollutants. In this connection, designing a hybrid adsorbent to remove a wide range of micropollutants can be potential area of future research [32]. Combination of inorganic and organic hybrid adsorbent such as using chitosan and copper can be considered as a promising candidate for efficient phosphate removal [51].

Material like graphene aerogel can be mixed with porous structure such as Fe_2O_3 nanocubes which recorded to have fast adsorption rate and outstanding reusability [52]. Porous material such as zeolite can be mixed with organic material such as chitosan. These two materials can either be composited into a film[53] and fine powder[54]. Result showed an improvement on adsorption of phosphate from using zeolite to composited zeolite with chitosan [55].

4. Sewage recycling

The five main types of use of recycled water indicated by Salgot et al. 2018 [56], which are irrigation, landscaping, water bodies recovery, groundwater recharge, drinking water supply as well as to be used in industry as cooling, in process and cleaning the machinery. Nowadays, applications of recycle water in vehicle washing, fire protection and air conditioning are being practiced [57]. However, reclaimed water being used as spray irrigation, cooling towers and toilet flushing are reported to cause potential fecal pathogen-associated health risks [58].

Water reclamation and reuse must follow the standards and recommendations issued at different administrative levels. As in China, Ministry of Construction and Standardization Administration issued a series of regulations and standards such as Reuse of Urban Recycling Water–Water Quality Standard for Urban Miscellaneous Water Consumption (GB/T 18920-2002) [59]. The majority of the suggested treatments are biological and such technologies can be efficient only at small sewage load[57]. In addition, the less contaminated sewage source e.g. kitchen wastewater is more favorable to be recycled compared to waste from toilet due to less process required prior to recycle [60]. The usage of recycle water is still rare and further studies are required to deeply understand the water resource and usage target[61].

5. Conclusion

Biological processes such as bioaugmentation and MBBR have relatively same limitations such as dependent on the temperature, types of selected strains and affected by toxic components. Membrane bioreactor is important for high quality of treated sewage. Due to its availability, development of adsorbent from waste material with high adsorption capacity and fast adsorption rate would be important in the future. Hybrid adsorbents are reported to have better performance compared to single adsorbent. If carried out properly, sewage recycling can be used as a sustainable solution for the benefit of mankind.

Acknowledgments

The authors would like to thank Universiti Malaysia Sabah for funding this project under UMS grant (SDK 0037 – 2018).

References

[1] Strategic Communication Department 2017 *Malaysia environmental quality report 2017* (Putrajaya, Malaysia: Enviroment Knowledge Center)

[2] Ashraf A, Saion E, Gharibshashi E, Kamari HM, YC Kong, MS Hamzah and MS Elias 2016 Rare earth elements in core marine sediments of coastal East Malaysia by instrumental neutron activation analysis *Appl. Radiat. Isot.* **107** 17–23

[3] Praveena SM, Shaifuddin SNM, Sukiman S, Nasir FAM, Hanafi Z, Kamarudin N, Ismail THT and Aris AZ 2018 Pharmaceuticals residues in selected tropical surface water bodies from Selangor (Malaysia): Occurrence and potential risk assessments *Sci. Total Environ.* **642** 230–40

[4] Goswami L, Vinoth Kumar R, Borah SN, Arul Manikandan N, Pakshirajan K and Pugazhenthi G 2018 Membrane bioreactor and integrated membrane bioreactor systems for micropollutant removal from wastewater: A review *J. Water Process Eng.* **26** 314–28

[5] Schaeffer J, Treguier C, Piquet JC, Gachelin S, Laureau NC, Saux JCL, Garry P and Guyader FSL 2018 Improving the efficacy of sewage treatment decreases norovirus contamination in oysters *Int. J. Food Microbiol.* **286** 1–5

[6] Stazi V and Tomei MC 2018 Enhancing anaerobic treatment of domestic wastewater: State of the art, innovative technologies and future perspectives *Sci. Total Environ.* **635** 78–91

[7] Liao BQ, Kraemer JT, and Bagley DM 2006 Anaerobic Membrane Bioreactors: Applications and Research Directions *Crit. Rev. Environ. Sci. Technol.* **36** 489–530

[8] Zarasvand Asadi R, Suja F, Ruslan MH, and Jalil NA 2013 The application of a solar still in domestic and industrial wastewater treatment *Sol. Energy* **93** 63–71

[9] Zulaikha S, Lau WJ, Ismail AF, and Jaafar J 2014 Treatment of restaurant wastewater using ultrafiltration and nanofiltration membranes *J. Water Process Eng.* **2** 58–62

[10] Tandukar M, Uemura S, Ohashi A, and Harada H 2006 Combining UASB and the 'fourth generation' down-flow hanging sponge reactor for municipal wastewater treatment *Water Sci. Technol.***53** 209–18

[11] Muserere ST, Hoko Z, and Nhapi I 2014 Characterisation of raw sewage and performance assessment of primary settling tanks at Firle Sewage Treatment Works, Harare, Zimbabwe *Phys. Chem. Earth* **67–9** 226–35

[12] Sarti A, Garcia ML, Zaiat M and Foresti E 2007 Study on Sewage Quality from Sewage Treatment Plant at Vashi, Navi Mumbai *Bioresour. Technol.* **5** 1508 - 13

[13] Wąsik E and Chmielowski K 2017 Ammonia and indicator bacteria removal from domestic sewage in a vertical flow filter filled with plastic material *Ecol. Eng.***106** 378–84

[14] Kulkarni B, Wanjule RV and Shinde HH 2018 Study on Sewage Quality from Sewage Treatment Plant at Vashi, Navi Mumbai *Mater. Today Proc.* **5** 1859–63

[15] Al-Shayah M and Mahmoud N 2008 Start-up of an UASB-septic tank for community on-site treatment of strong domestic sewage *Bioresour. Technol.* **99** 7758–66

[16] Herrero M and Stuckey DC 2015 Bioaugmentation and its application in wastewater treatment: A review *Chemosphere* **140** 119–28

[17] Tang HL and Chen H 2015 Nitrification at full-scale municipal wastewater treatment plants: Evaluation of inhibition and bioaugmentation of nitrifiers *Bioresour. Technol.* **190**76–81
[18] Han L, Pei H, Hu W, Jiang L, Ma G and Han F 2015 Integrated campus sewage treatment and biomass production by Scenedesmus quadricauda SDEC-13 *Bioresour. Technol.* **175** 262–68
[19] Hesnawi R, Dahmani K, Al-Swayah A, Mohamed S, and Mohammed SA 2014 Biodegradation of Municipal Wastewater with Local and Commercial Bacteria *Procedia Eng.* **70** 810–14
[20] Talaiekhozani A and Rezania S 2017 Application of photosynthetic bacteria for removal of heavy metals, macro-pollutants and dye from wastewater: A review *J. Water Process Eng.* **19** 312–21
[21] Huang C, Luo MT, Chen XF, Xiong L, Li XM, and Chen XD 2017 Recent advances and industrial viewpoint for biological treatment of wastewaters by oleaginous microorganisms *Bioresour. Technol.* **232** 398–407
[22] Benner J *et al. 2013* Is biological treatment a viable alternative for micropollutant removal in drinking water treatment processes? *Water Res.* **47** 5955–76
[23] Kermani M, Bina B, Movahedian H, Amin MM and Nikaein M 2008 Application of moving bed biofilm process for biological organics and nutrients removal from municipal wastewater *Am. J. Environ. Sci.* **4** 675–82
[24] Zinatizadeh AAL and Ghaytooli E 2015 Simultaneous nitrogen and carbon removal from wastewater at different operating conditions in a moving bed biofilm reactor (MBBR): Process modeling and optimization *J. Taiwan Inst. Chem. Eng.* **3** 98–111
[25] Boltz JP, Smets BF, Rittmann BE, van Loosdrecht MCM, Morgenroth E and Daigger GT 2017 From biofilm ecology to reactors: a focused review *Water Sci. Technol.* **75** no. 8 1753–60
[26] Rusten B, Eikebrokk B Ulgenes Y and Lygren E 2006 Design and operations of the Kaldnes moving bed biofilm reactors *Aquac. Eng.* **34** 322–31
[27] Baban A and Talinli I 2009 Modeling of organic matter removal and nitrification in sewer systems - an approach to wastewater treatment *Desalination* **246** 640–47
[28] Zhang X, Chen X, Zhang C, Wen H, Guo W and Ngo H 2016 Effect of filling fraction on the performance of sponge-based moving bed biofilm reactor *Bioresour. Technol.* **219** 762–67
[29] Chu L and Wang J 2011 Nitrogen removal using biodegradable polymers as carbon source and biofilm carriers in a moving bed biofilm reactor *Chem. Eng. J.* **170** 220–25
[30] Martín-Pascual J, López-López C, Cerdá A, González-López J, Hontoria E and Poyatos JM 2012 Comparative kinetic study of carrier type in a moving bed system applied to organic matter removal in urban wastewater treatment *Water. Air. Soil Pollut.* **223** 1699–1712
[31] Brepols C 2011 *Operating Large Scale Membrane Bioreactors for Municipal Wastewater Treatment* 2011 (London, United Kingdom: IWA Publishing) p9
[32] Khan MA, Ngo HH, Guo W, Liu Y, Chang SW, Nguyen DD, Nghiem LD and Liang H 2018 Can membrane bioreactor be a smart option for water treatment? *Bioresour. Technol. Reports* **4** 80–7
[33] Zhou L, Ye B and Xia s 2017 Assessing membrane biofouling and its gel layer of anoxic/oxic membrane bioreactor for megacity municipal wastewater treatment during plum rain season in Yangtze River Delta, China *Water Res.* 127 22–31
[34] Buzatu P, Qiblawey H, Odai A, Jamaleddin J, Nasser M and SJ Judd 2018 Clogging vs. fouling in immersed membrane bioreactors *Water Res.* **144** 46–54
[35] Liang S, Qi G, Xiao K, Sun J Giannelis EP, Huang X, Elimelech M 2014 Organic fouling behavior of superhydrophilic polyvinylidene fluoride (PVDF) ultrafiltration membranes functionalized with surface-tailored nanoparticles: Implications for organic fouling in membrane bioreactors *J. Memb. Sci.* **463** 94–101
[36] Liu R, Huang X, Sun YF and Qian Y 2003 Hydrodynamic effect on sludge accumulation over membrane surfaces in a submerged membrane bioreactor *Process Biochem.* **39** 157–163
[37] Yan X, Xiao K, Liang S, Lei T, Liang P, Xue T, Yu K, Guan J and Huang X 2015 Hydraulic optimization of membrane bioreactor via baffle modification using computational fluid dynamics *Bioresour. Technol.* **175** 633–37
[38] Yan X, Xiao K, Liang S, Lei T, Liang P, Xue T, Yu K, Guan J and Huang X 2015

Hydrodynamic optimization of membrane bioreactor by horizontal geometry modification using computational fluid dynamics *Bioresour. Technol.* **200** 328–34
[39] Rosenberger S, Krüger U, Witzig R, Manz W, Szewzyk U and Kraume M 2002 Performance of a bioreactor with submerged membranes for aerobic treatment of municipal waste water *Water Res.* **36** 413–20
[40] Wang H, Chen Z, Miao J and Li Y 2016 A novel approach for mitigation of membrane fouling: Concomitant use of flocculant and magnetic powder *Bioresour. Technol.* **209** 318–25
[41] Cashman S, Ma X, Mosley J, Garland J, Crone B and Xue X 2018 Energy and greenhouse gas life cycle assessment and cost analysis of aerobic and anaerobic membrane bioreactor systems: Influence of scale, population density, climate, and methane recovery *Bioresour. Technol.* **254** 56–66
[42] Liudvinaviciute D, Almonaityte K, Rutkaite R, Bendoraitiene J and Klimaviciute R 2018 Adsorption of rosmarinic acid from aqueous solution on chitosan powder *Int. J. Biol. Macromol.***118** 1013–20
[43] Huang D, Wu J, Wang L, Liu X, Tang X, Tang C and Xu J 2018 Novel insight into adsorption and co-adsorption of heavy metal ions and an organic pollutant by magnetic graphene anomaterials in water *Chem. Eng. J.*
[44] Low SK, Tan MC, and Chin NL 2018 Effect of ultrasound pre-treatment on adsorbent in dye adsorption compared with ultrasound simultaneous adsorption *Ultrason. Sonochem.* **48** 64–70
[45] Ranade VM and Bhandari VV 2014 *Industrial Wastewater Treatment, Recycling and Reuse*. (Oxford, UK: Elsevier Ltd)
[46] Meng X *et al.* 2018 Comparison of the reduction of chemical oxygen demand in wastewater from mineral processing using the coagulation–flocculation, adsorption and Fenton processes *Miner. Eng.* **128** 275–83
[47] Metcalf and Eddy 2003 *Wastewater Engineering : Treatment and Reuse* ed 4 (Boston: McGraw-Hill)
[48] Kim J, Hwang MJ, Lee SJ, Noh W, Kwon JM, Choi JS and Kang CM 2016 Efficient recovery of nitrate and phosphate from wastewater by an amine-grafted adsorbent for cyanobacterial biomass production *Bioresour. Technol.* **205** 269–73
[49] Zahrim AY, Lija Y Ricky LNS and Azreen I 2016 Fruit waste adsorbent for ammonia nitrogen removal from synthetic solution: Isotherms and kinetics *IOP Conf. Ser. Earth Environ. Sci.* vol 36
[50] Azreen I, Lija Y and Zahrim AY 2017 Ammonia nitrogen removal from aqueous solution by local agricultural wastes *IOP Conf. Ser. Mater. Sci. Eng.* vol 206 p 12077
[51] An B, Lee S, Kim HG, Zhao D, Park JA and Choi JW 2019 Organic/inorganic hybrid adsorbent for efficient phosphate removal from a reservoir affected by algae bloom *J. Ind. Eng. Chem.* **69** 211–16
[52] Yu X, Wei Y, Liu C, Ma J, Liu H, Wei S, Deng W, Xiang J and Luo S 2019 Ultrafast and deep removal of arsenic in high-concentration wastewater: A superior bulk adsorbent of porous Fe_2O_3 nanocubes-impregnated graphene aerogel *Chemosphere* **222** p. 258–66
[53] Kołodyńska D, Hałas P, Franus M and Hubicki Z 2017 Zeolite properties improvement by chitosan modification—Sorption studies *J. Ind. Eng. Chem.* **52** p.187–96
[54] Kumar S, Prasad K, Gil JM, Sobral AJFN and Koh J 2018 Mesoporous zeolite-chitosan composite for enhanced capture and catalytic activity in chemical fixation of CO_2 *Carbohydr. Polym.***198** 401–06
[55] Xie J, Li C, Chi L and Wu D 2013 hitosan modified zeolite as a versatile adsorbent for the removal of different pollutants from water *Fuel* **103** 480–85
[56] Salgot M, Folch M and Unit SS 2018 astewater treatment and water reuse *Curr. Opin. Environ. Sci. Heal.* **2** 64–74
[57] Pidou M, Avery L, Stephenson T, Jeffery P, Parsons SA, Liu S, Memon FA and Jefferson B 2008 Chemical solutions for greywater recycling *Chemosphere* **71** 147–55
[58] Hamilton KA, Hamilton MT, Johnson W, Jjemba P, Bukhari Z, LeChevallier M and Haas CN

2018 Health risks from exposure to Legionella in reclaimed water aerosols: Toilet flushing, spray irrigation, and cooling towers *Water Res.* **134** 261–79
[59] Lyu S, Chen W, Zhang W, Fan Y and Jiao W 2016 Wastewater reclamation and reuse in China: Opportunities and challenges *J. Environ. Sci.* **39** 86–96
[60] Domènech L, March H, Vallès M and Saurí D 2015 Learning processes during regime shifts: Empirical evidence from the diffusion of greywater recycling in Spain *Environ. Innov. Soc. Transitions* **15** 26–41
[61] Lu W and Leung AYT 2003 A preliminary study on potential of developing shower/laundry wastewater reclamation and reuse system *Chemosphere* **52** 1451–59

IOP Conf. Series: Materials Science and Engineering **606** (2019) 012008 doi:10.1088/1757-899X/606/1/012008

Observation on the drag reducing effect of low concentration chitosan solution in turbulent pipe flow

M. A. Asidin[1], E. Suali[1*], T. Jusnukin[1], F.A. Lahin[1]
[1]Faculty of Engineering, UMS, Kota Kinabalu, Sabah, Malaysia
Email: *emma.suali@ums.edu.my*

Abstract. Polymer-based drag reducing agent (DRA) is used widely in various industries due to its ability to enhance fluid flow as well as economically beneficial to various sectors. This study is conducted to observe the drag reducing effect of chitosan as DRA and observed its effect on the turbulent flow inside the pipe. The experiment is conducted in a closed loop circulation system where water is the transport medium. The pipe system consist of acrylic pipes with 0.013 m, 0.025 m and 0.038 m diameter. The chitosan was extracted from shrimp shells and five different concentration of chitosan were tested. The degree of deacetylation of the chitosan shows 97% which indicates its solubility in water. It was found that the highest drag reduction percentage were obtained from 0.038 m pipe of 33 ppm concentration which is 28%. The flow pattern inside the pipe is also analysed and discussed in this study

1. Introduction

Turbulence inside a pipeline caused fluid particles to move randomly which eventually lead to friction. Drag is defined as the frictional force that acts against the motion of an object through a fluid [1]. Pressure head losses happened due to drag inside the pipeline and cause the flow rate of the fluid to decrease. Thus, more energy is needed to uphold the desired flow rate [2] which result in installing more pumping station. Therefore, many methods have been researched to tackle this problem throughout the years and known as drag reduction. Abdulbari et al. [3] defined drag reduction as the science of improving fluid flow by reducing the pressure drop across a channel. The most well-known method to reduce drag is by using additives. These additives are known as Drag Reducing Agent (DRA) and consist of three types: polymer, surfactant, and solid suspensions. DRAs are very viscoelastic and mostly hydrocarbon which does not affect the physical properties of the transported fluid. The mixture between DRA and solvent will produce a solution that is shear-degradable, viscoelastic, time-independent and non-Newtonian fluids [4]. Polymeric DRA is considered as the most effective and economically beneficial due to its rheological properties and its resistance to shear force [3]. Polymeric DRAs are the long-chain polymer with high molecular weight. The polymer used commercially nowadays have a fast dissolution rate and slow degradation rate [5]. When injected into a pipeline, the polymer DRA will interact with the fluid flow and dampened the outbreak of turbulence namely at the buffer region of the flow. Lumley [6] proposed that the stretching of a coiled polymer will increase the extensional viscosity and dampened the small eddies in the buffer layer resulting in a more effective drag reduction. Polymer DRA is widely applied in various sectors such as firefighting [7], district cooling and heating system [8,9], sewage system [10] and also medical field [11,12]. In spite of the benefit of polymer in drag reduction, there are a few major issues regarding the use of artificial polymers. The effectiveness of polymers will reduce in time due to mechanical degradation. Early experiments on mechanical degradation stated that the degradation mostly occurs due to its dependence to molecular weight [13] and the chain scission that occurs around the midpoint of the polymer chain suggesting that the polymer is highly extended when it breaks [14]. These theories are further supported by Shanshool et al. [15] which stated that high molecular weight polymers prone to shear-induced degradation, and polymers with linear -chain structure are more vulnerable than branched polymers. In recent years, there are some concerns about the toxicity and environmental impact of conventional polymer additives [3]. Despite the proven excellence in drag reduction properties, these synthetic polymers will affect the environment

IOP Conf. Series: Materials Science and Engineering **606** (2019) 012008 doi:10.1088/1757-899X/606/1/012008

due to the chemical contents build up in the ground and become pollutants [16]. Therefore, researchers have begun to switch to new natural and organic additives as alternatives to conventional DRA. This research was conducted to study the potential of an environmentally friendly DRA extracted from biopolymers and observe its ability to reduce drag at low concentration and its effect to different pipe diameter. The flow pattern inside the pipeline is observed and discussed. The DRA for this research is chitosan which is extracted from shrimp shell. These shrimp shells consist of 20-30% of chitin, 30-40% of protein, and 30-50% of calcium [17]. Chitosan can be a used as DRA due to its abundance in nature and its non-toxicity, biocompatibility, biodegradability, and adsorption.[18].

2. Materials and methods

2.1. Preparation of chitosan

The raw material of fresh shrimp shells was dried up to produce crispy shells. The shells were grounded to form a powder. The powder was demineralized using 5% HCl for 24 hours at room temperature. The ratio of the weight of shells and volume of HCl (w/v) is 1:5 [19]. After that, it is rinsed with water to neutralize the pH value. It was then treated with 10% HCl to ensure complete demineralization of the shell which showed no bubble formation. Then, the powder was deproteinized using 5% NaOH and dried up to produce chitin. Finally, the chitin was deacetylated by treating it into 60% of NaOH solution. The solution is heated then rinsed with distilled water and dried up to produce chitosan. The chitosan powder produced was mixed with distilled water and stirred constantly until the solution fully mixed and become homogenous. During the mixing processed, heat is constantly supplied to prevent clumping. This solution must be continuously stirred because the polymer is insoluble in water. Then, the solution is cooled down at ambient temperature before the addition of 100% glacial solutions of acetic acids. Different concentrations of chitosan solution which are 7 ppm, 13 ppm, 20 ppm, 27 ppm, and 33 ppm were prepared based on the amounts of acetic acid solution added. The solution needs to be mixed with water before injecting into the pipe systems a night before to ensure its solubility in water.

2.2. Experimental rig

The rig is a closed loop circulatory system with water as the transporting medium. The design of the experimental rig is shown in figure 1. The system consists of reservoir tanks, pipes, valves, and pumps. The main reservoir tank is joined by two pipes which connect the centrifugal pump with the piping system. The centrifugal pump will deliver water across the system. This pipe design consists of three transparent polyvinyl chloride (PVC) pipe that has different size of inside diameter (0.013 m, 0.025 m, and 0.038 m). These pipes have testing sections where all the readings will be taken during the experiment. The main purpose of the rig is to study the drag reducing effect of chitosan with low concentrations in different pipe diameters and flowrates.

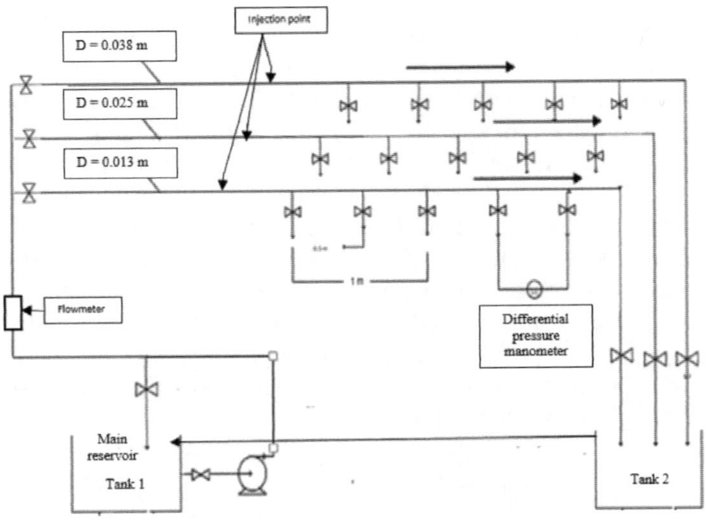

Figure 1: Schematic diagram of the experimental rig.

Flow piping system started from the main reservoir tank. The water is pumped until it reaches the connection that split the pipe into three sections with different pipe diameter and testing section. The flowrate is controlled and determined using the flowmeter connected to the main pipe. The experiment is conducted at two different flowrate which are 0.5 m^3/h and 1 m^3/h. The pressure drop data before the addition of additives which is the pressure drop reading of raw water flow will be recorded first before the DRA is added. The DRA is added into the system from the injection point by using a peristaltic pump. There are five variables of chitosan concentration varied from 7 ppm to 33 ppm. The pressure drop value can be read at the moment the DRA solution has begun to flows across the test section. The pressure drop reading is obtained by using the differential pressure manometer. Since the pipes at the testing sections are transparent, the flow pattern of the solution can be observed by looking at the visible pipe wall. The observation of the flow pattern is important to analyze the interaction between the water medium and the DRA. It will be conducted by using a high definition camera. The solutions that have flown across the pipeline will then reach the reservoir tank 2 and will flow back into the main reservoir tank and this process is continuous.

3. Results and discussion

The performance of chitosan as DRA is determined by the drag reduction percentage. The drag reduction percentage of pipe flow is obtained based on pressure drop or head loss before and after the addition of DRA into the water circulation system as expressed by the following equation:

$$\%DR = \frac{\Delta P - \Delta P_{dra}}{\Delta P} \; x \; 100 \qquad (1)$$

Where ΔP is the value of the head loss of tap water before the addition of chemical additive and ΔP_{DRA} is the value of pressure drop after addition of DRA into the system.

IOP Conf. Series: Materials Science and Engineering **606** (2019) 012008 doi:10.1088/1757-899X/606/1/012008

3.1. Effect of different DRA concentration

The percentage of drag reduction against chitosan concentration for flowrate 1 m^3/h is shown in Figures 2. Generally, the graph shows an increasing trend in drag reduction. The graph also shows that the percentage of drag reduction increase as the concentration of chitosan increase for each pipe size. For all pipe size, the highest drag reduction percentage obtained is 28% for 0.038 m, 29% for 0.025 m and 6% for 0.013 m. The %DR increase notably in 0.025 m pipe with the concentration of 21 ppm to 33 ppm. This result agrees with the early observation made by Tom in 1948 which stated that only a small amount of additives needed to reduce drag significantly. In Figure 3, there is a slight fluctuation of drag reduction percentage for pipe diameter 0.025 m and 0.013 m. The %DR decrease at 21 ppm concentration and increase at 27 ppm and 33 ppm. From the graphs, we can propose that a bigger diameter pipe will produce higher drag reduction compared to a pipe with smaller diameter. Several researchers have investigated the effects of internal diameter to drag reduction performance and found that the %DR increase as the diameter increase [20,21]. Karami and Mowla [22] investigating the %DR of three different solutions in two different pipe diameters and observed that DR decreases as the pipe diameter increase which is contradicting with the result of this research. Theoretically, this phenomena is related to the eddied formation inside the turbulent flow and the capabilities of the DRA to suppress that formation. At larger pipe diameter, the magnitude of eddies is smaller but in a larger size. Thus, they can be broken and suppress easily by the DRA. Compared to smaller pipe diameter, even though these eddies formed are small, they are higher in number. The interaction between the DRA and the fluid takes more time to break up most of the eddied before the energy is lost [23].

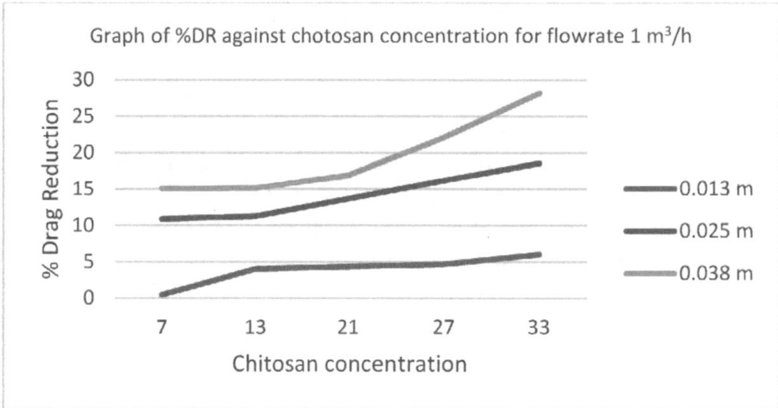

Figure 2: Graph of %DR against chitosan concentration for flowrate 1 m^3/h

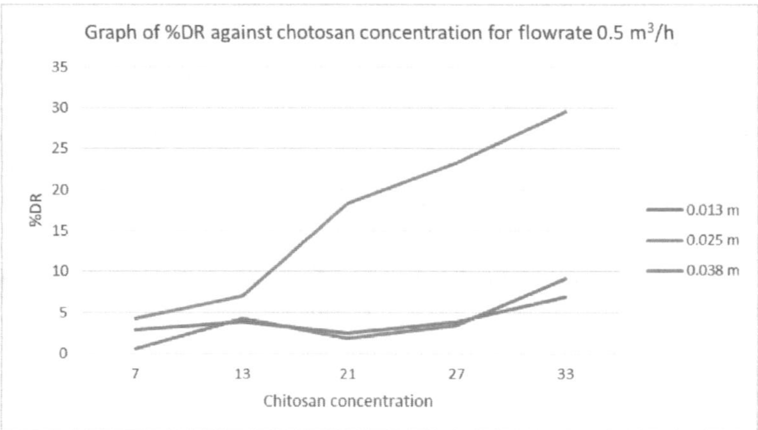

Figure 3: Graph of %DR against chitosan concentration for flowrate 0.5 m^3/h

3.2. Observation od flow pattern inside the pipe

The observation of flow pattern inside the pipe is made using a high definition camera. The visualization is made on the 0.038 m diameter pipe because there are no visible differences obtained from the pipes with 0.013 m and 0.025 m diameter. The parameters for this observation are kept constant which is 1 m^3/h flowrate and 33 ppm chitosan concentration. The parameters are constant to compare the flow pattern of the fluid with and without additives. Figure 4 shows the visualization of the flow inside the pipe that shows the difference between pipe flow without and with the addition of DRA. There are some fluctuation near the pipe wall in Figure 4(a) indicating that there are some eddies forming. These eddies are caused by the chaotic movement of the fluid particle inside the pipe. Eddies formed are randomly distributed alongside the pipe wall. The flow in Figure 4(b) shows more stream-wise fluctuations near the pipe wall. The flow after the addition of DRA is more streamline and define compared to the flow without DRA. The difference that can be seen between these two types of flow is that the flow with DRA appears to be as horizontal stripes [24].

Figure 4(a): The flow inside the pipe without DRA.

Figure 4(b): The flow inside the pipe with DRA.

Figure 4: The visualization of the flow inside the pipe that shows the difference between pipe flow without and with the addition of DRA.

4. Conclusion

Chitosan is a biopolymer with polysaccharides properties such as non-toxicity and biodegradability. The process involves in the extraction of chitosan from shrimp shell is demineralization, deproteinization, deacetylation, and dilution. Chitosan shows good drag reduction ability with the highest %DR is 28% for 33 ppm chitosan concentration in 0.038 m pipe diameter and 1 m^3/h flowrate. The drag reduction percentage increase as the chitosan concentration increase. The diameter of the pipe will affect the drag reduction performance. Bigger pipe diameter produces a smaller amount of eddies compared to smaller pipe diameter but bigger in size. This will give the chitosan enough time to interact and suppress eddies formation. This interaction makes the flow inside the pipe more streamline and lower the pressure drop.

5. References

[1] S.S. Salehudin, S. Ridha, Coconut Residue as Biopolymer Drag Reducer Agent in Water Injection System, *Int. J. Appl. Eng. Res.* **11** (2016) 8037–8040.

[2] Q. Muslim, A. Ali, Drag Force Reduction of Flowing Crude Oil by Polymers Addition, *Iraqi J. Mech. Mater. Eng.* **8** (2008) 149–161.

[3] H.A. Abdulbari, A. Shabirin, H.N. Abdurrahman, Bio-polymers for improving liquid flow in pipelines-A review and future work opportunities, *J. Ind. Eng. Chem.* **20** (2014) 1157–1170.

[4] R. Martínez-Palou, M. de L. Mosqueira, B. Zapata-Rendón, E. Mar-Juárez, C. Bernal-Huicochea, J. de la Cruz Clavel-López, J. Aburto, Transportation of heavy and extra-heavy crude oil by pipeline: A review, *J. Pet. Sci. Eng.* **75** (2011) 274–282.

[5] B.A. Jubran, Y.H. Zurigat, M.F.A. Goosen, Drag Reducing Agents in Multiphase Flow Pipelines: Recent Trends and Future Needs, *Pet. Sci. Technol.* **23** (2005) 1403–1424.

[6] J.L. Lumley, Drag reduction in two phase and polymer flows., *Phys. Fluids*. **20** (1977).

[7] R.C.R. Figueredo, E. Sabadini, Firefighting foam stability: The effect of the drag reducer poly(ethylene) oxide, Colloids Surfaces A Physicochem. *Eng. Asp*. **215** (2003) 77–86.

[8] M.M.A. El-azm, S.Z. Kassab, S.A. Elshafie, Experimental and Numerical Study for Turbulent Flow Drag Reduction in District Cooling Systems, **6** (2014) 113–125.

[9] N.J. Kim, S. Kim, S.H. Lim, K. Chen, W. Chun, Measurement of drag reduction in polymer added turbulent flow, Int. Commun. Heat Mass Transf. 36 (2009) 1014–1019.

[10] R.H.J. Sellin, Drag Reduction in Sewers: First Results From a Permanent Installation, *J. Hydraul. Res.* **16** (1978) 357–371.

[11] J.N. Marhefka, M. V Kameneva, Natural Drag-Reducing Polymers: Discovery, Characterization and Potential Clinical Applications, (2000) 1–12.

[12] J.N. Marhefka, P.J. Marascalco, T.M. Chapman, A.J. Russell, M. V Kameneva, Poly(N-vinylformamide)A Drag-Reducing Polymer for Biomedical Applications, Biomacromolecules. **7** (2006) 1597–1603.

[13] J.D. Culter, J.L. Zakin, G.K. Patterson, Mechanical degradation of dilute solutions of high polymers in capillary tube flow, *J. Appl. Polym. Sci.* **19** (1975) 3235–3240.

[14] A.F. Horn, E.W. Merrill, Midpoint Scission of Macromolecules in Dilute solution in turbulent flow, *Nat. Publ.* **312** (1984).

[15] J. Shanshool, M. F.A., I.N. Slaiman, The influnace of mechanical effects on degradation of polyisobutylene as drag reducing agent, *Pet. Coal.* **53** (2011) 218–222.

[16] H. Kaur, A.P.G. Singh, A. Jaafar, U.T. Petronas, The Study of Drag Reduction Ability of Naturally Produced Polymers from Local Plant Source, in: *Int. Pet. Technol. Conf.*, 2013.

[17] H.A.A. Bari, N.K. Mohamad, N. Mohd, A.H. Nour, Effect of chitosan solution in turbulent drag reduction in aqueous media flow, **6** (2011) 3058–3064.

[18] S.M. Hudson, C. Smith, Biopolymers from Renewable Resources, 1998.

[19] N. Van Toan, Production of Chitin and Chitosan from Partially Autolyzed Shrimp Shell Materials, *Open Biomater. J.* **1** (2009) 21–24.

[20] P.S. Virk, Drag reduction fundamentals, AIChE J. **21** (1975) 625–656.

[21] A.A. Khadom, A.A. Abdul-Hadi, Performance of polyacrylamide as drag reduction polymer of crude petroleum flow, *Ain Shams Eng. J.* **5** (2014) 861–865.

[22] H.R. Karami, D. Mowla, Investigation of the effects of various parameters on pressure drop reduction in crude oil pipelines by drag reducing agents, *J. Nonnewton. Fluid Mech.* **177–178** (2012) 37–45.

[23] H.A. Abdul Bari, E. Suali, Z. Hassan, Glycolic Acid Ethoxylate Lauryl Ether Performance as Drag Reducing Agent in Aqueous Media Flow in Pipelines, *J. Appl. Sci.* **8** (2008) 4410–4415.

[24] W. Gu, Y. Kawaguchi, D. Wang, Experimental Study of Turbulence Transport in a Dilute

Surfactant Solution Flow Investigated by PIV, **132** (2010) 1–7.

Acknowledgments
This work was supported by the grant FRG0416-TK-1/2015 from the Ministry of Higher Education Malaysia and grant GUG0052-TK-2/2016 from UMSgreat. Sincere gratitude to fellow corresponding authors for guidance and persistent support.

IOP Conf. Series: Materials Science and Engineering **606** (2019) 012009 doi:10.1088/1757-899X/606/1/012009

Optimisation of parameters for detection of manganese ion using electrochemical method

R Gilbert[1], S Siddiquee[1*], S Saallah[1] and Tamrin M Lal[2]

[1] Biotechnology Research Institute, University Malaysia Sabah, Jalan UMS, 88400 Kota Kinabalu, Sabah, Malaysia

[2] Borneo Research Marine Institute, University Malaysia Sabah, Jalan UMS, 88400 Kota Kinabalu, Sabah, Malaysia

Email: shafiqpab@ums.edu.my

Abstract. An electrochemical method was detected a heavy metal ion of manganese (Mn^{2+}) based on bare gold electrode (AuE). The bare AuE was optimized including various type of electrolytes, redox indicators, pH, scan rate and accumulation time using cyclic voltammetry (CV). From the CV results, the optimal conditions of Mn^{2+} were found of Tris-HCl (0.1 M, pH 2) buffer for electrolytes supported by 5 mM Prussian blue as a redox indicator with scan rate of 100 mVs^{-1} and accumulation time of 2 s, respectively. Differential pulse voltammetry (DPV) method was detected with different concentrations of Mn^{2+} on the ranged from 0.05 to 1.0 ppm with limit of detection (LOD) of 0.075 mg L^{-1} (S/N=3). A linear regression was found of 0.9659 (r^2). The above results indicated that the bare AuE was unable to detect of Mn^{2+} below safety level. So, the proposed method suggests to detect below safety level of Mn^{2+} based on the further modification of AuE with high sensitivity, selectivity and accuracy.

1. Introduction

Manganese is the third most abundant transition element in the earth's crust as compared to other elements. Manganese plays important role as a key activation of many enzymes in the human, animals and plants [1,2]. The manganese's trace is important in cellular metabolism [3]. Manganese has been applied on manufacture industries such as batteries, fertilizers, ceramics, dietary supplements and electrical coils [1,2]. The high concentrations of manganese can cause adverse negative effects to human health, environment and the ecosystem [1]. The high level of manganese exposure which linked to neurological disorders such as Parkinson-like syndrome. Victim will suffer the manganism diseases which physical disability and body posture problems due to the low muscle stretching and flexibility [4]. In addition, Manganism also affecting the brain activities which resulted in decreased mental status, slow speech, hard to adapt new environment and lack of interest [5]. The dissolved manganese ion widely present in drinking water, foods, ad plants [6].

Due to it existent, the standard quality levels of manganese in drinking water have been fixed for domestic water was 0.05 $\mu g\ mL^{-1}$ [3,7]. Note that, the other journal mentioned the different value on maximum concentration allowance for domestic water was 0.05 mg mL^{-1} [2]. The Ministry of Heath

Malaysia has been set the standard quality level for manganese in drinking water was 0.1 mg L^{-1} [8]. Despite on the usages of domestic water, the detection of manganese ion becomes highly important in order to monitor the uncertainty amount of manganese in domestic water such as drinking water.

Currently, various conventional methods had been used such energy dispersive X-ray florescence spectroscopy (EDXRF) [9], atomic absorption spectrometry (AES) [10,11], microwave plasma-atomic emission spectrometry (MP-AES) [12] and coupled plasma mass spectrometry (ICP-MS) [13]. These methods provided high selectivity and sensitivity for manganese ion detection but suffer from a number of drawbacks including expensive instruments, tedious, complex and time-consuming procedures, and require skilled operators. On the other hand, electrochemical techniques offer a practical alternative and have drawn growing attentions from researchers due to the advantages offered such as high selectivity and sensitivity, rapid with quality and quantity analysis, on-site detection (portable), inexpensive instrument and does not required pre-treatment of sample before analysis was conducted [7,14,15,16]. Recently, less information on detection of manganese using electrochemical methods. Latest researches only done by Roushani et al. (2017) and Ruggeri et al. (2017) which both were using glassy carbon electrode (GCE)[1,6]. No one has use the gold electrode on detection of manganese ion yet. However, the gold electrode has proven with high conductivity, electrocatalytic, good selectivity and high signal-to-noise ratio. The utilization of gold electrode in electrochemical detection of various elements have been widely reported. According to Jiang et al. (2016), the detection of arsenic (As^{3+}) using gold electrode fabricated with ultrathin SnO2 nanosheets was showed the limit of detection about 5 mg L^{-1} in 0.1 M acetate buffer (pH 5)[17]. Another experiment conducted by Giacomino et al. (2011) on the analytical performance using lateral gold electrode showed the limit of detection As^{3+} about 0.06 mg L^{-1}[18].

As this research is ongoing research, many parameters have been considered and optimized. The electrochemical parameters such as type electrolytic solution, redox indicator, pH, scan rate and accumulation time had been optimized using cyclic voltammetry (CV) method. The optimal parameters were used for measured the different levels of manganese towards current signals using differential pulse voltammetry (DPV) method. The proposed sensor enhances the high sensitivity and selectivity with focusing on optimization step.

2. Experimental section

2.1 Materials and chemicals

Potassium hydrogen phosphate (K_2HPO_4), potassium dihydrogen phosphate (KH_2PO_4), sodium citrate dihydrate ($C_6H_9Na_3O_9$) and sodium chloride (NaCl) were purchased from Systerm, citric acid anhydrous ($C_6H_8O_7$) was obtained from Nacalai Tesque, ammonium acetate ($C_2H_7NO_2$) was from Ajax Chemical and tris-HCl ($NH_2C(CH_2OH)_3$.HCl was from 1st Base for buffer preparation materials. Redox indicator chemicals, potassium hexacyanoferrate (III) and potassium ferrocyanide (II) trihydrate were the product of Nacalai Tesque, while methylene blue and iron (III) chloride were purchased from Systerm Chemicals and Sigma-Aldrich, respectively. Manganese (II) sulphate ($MnSO_4.H_2O$) as a target ion was obtained from Systerm.

2.2 Instrumentation

All electrochemical experiments were carried out with potentiostat/galvanostat (PGSAT) equipment (Metrohm-Autolab B.V), packaged with NOVA Autolab 1.8 software. The instrument uses three-electrode configuration systems which consist of working electrode (AuE), counter electrode (Pt wire) and reference electrode. The cyclic voltammetry pontensiostat indicate the presence of target ion through current responses, shown in cyclic voltammogram graph. Differential pulse voltammetry (DPV) determined the effect of different concentrations of target ion in the electrochemical cell by promoting the current changing in form of bell-shaped curve.

2.3 Pre-treatment of bare gold electrode (AuE)
Pre-treatment of the bare gold electrode (AuE) was performed following the method described by Siddiquee et al. (2010)[19]. Briefly, AuE was polished with 0.3 – 0.5 μm alumina slurry for 2 minutes. Then, the electrode was gently cleaned with distilled water before further cleaned with ultrasonic cleaner for 2 minutes. After that, the electrode was dried using nitrogen gas. The dried electrode was dropped with 5 μL of redox indicator at the centre of the electrode and left for 2 minutes [20].

2.4 Preparation of redox indicator
The Prussian blue as a redox indicator was prepared by mixing potassium ferrocyanide (II) trihydrate ($C_6H_6FeK_4N_6O_3$) and iron (III) chloride ($FeCl_3$) at 1:1 ratio [21]. 5 mM Prussian blue ($C_{18}Fe_7N_{18}$) was gently dropped on the surface of gold electrode. This redox indicator acts as supporting agent to promote high current response when it submerged into the electrolytic solution containing Mn^{2+}.

2.5 Preparation of electrolytic solution
Five types of electrolytic solution including phosphate buffer saline (PBS), ammonium, citrate, tris-HCl and acetate buffers were prepared in 10 ml with concentration of 0.1 M in the presence of target ion [22,23,24]. The tris-HCl buffer was selected as the optimal buffer due to the highest current response. The preparation of tris-HCl buffer was performed by mixing 0.1 M tris-HCl ($NH_2C(CH_2OH)_3$.HCl with 0.1 M sodium chloride (NaCl) in distilled water.

2.6 Analytical procedure
Data analysis were conducted using electrochemical sensor (potentiostat/galvanostat) machine. Total volume of the electrolytic solution containing the target ion was in the electrochemical cell is 10 mL [23,25]. The optimisation of buffer, redox indicator, pH, scan rate and accumulation time were performed using cyclic voltammetry (CV) method under electrochemical technique. Current responses from five type of buffers and three types of redox indicator with different pH were obtained to select the best buffer and redox indicator for further optimisation.

The initial electrochemical measurements of cyclic voltammetry method were 0.0 V to 1.7 V for potential applied; 0.8 V stop/start potential; 100 mVs^{-1} scan rate and 10 s accumulation time. First, each of the buffers was tested with different pH values ranging from 2 to 9. After pre-treatment of the AuE, the dried electrode was submerged into the 10 mL of 0.1 M electrolytic solution of different type of buffer containing 0.1 ppm of the manganese. Then, the AuE supported with different type of redox indicator (5 mM, 5 μL) was submerged into the selected electrolytic solution containing manganese for selection of the buffer and best redox indicator. The pH, buffer and redox indicator that yield the highest current response were then used for optimisation of the scan rate and accumulation time. The scan rate of the experiment was set from 50 – 350 mVs^{-1} with interval scan rate was 50 mVs^{-1} while the accumulation time was from 5 – 40 s with 5 s interval time. The optimised parameters obtained from the CV measurement were applied to DPV method to determine the effect of different concentrations to current signals. The linear range (LOR) of manganese was investigated from 0.05 to 1.0 ppm with 0.2 ppm interval of concentration. The peak formed at 0.67 V after 10 s deposition time. The concentration of manganese was inversely proportional with current response.

3. Results and discussion
3.1. Effect of pH
To test the effect of pH on detection of 0.1 ppm of manganese ion, we ranged the pH from pH 2 to pH 9 for the electrolytic solution. Adjustment of the pH was made by using 0.1 M of hydrochloric acid and 0.1 M of potassium hydroxide. From the result shown in Figure 1, the behaviours of Mn^{2+} under different pH of the electrolytic solution can be observed where the current responses increased with the decrease of pH of the electrolytic solution. For the oxidation, the complex mechanisms of Mn^{2+}

involved between the bulk and the solid (electrode) caused the peak to increase as the pH decreased. Thorough understanding of these mechanisms is still under debate. All the peaks were formed higher at oxidation range due to hydrogen evolution potential which more positive response compared to potential located at the reduction range of Mn^{2+} [26]. On the other hand, reduction peak was a good quantitative measurement to determine the concentration of Mn^{2+} in the solution [6]. This is due to the concentration of Mn^{2+} depending on the acidity of the solution as shown in Figure 1. Highest peak indicates the high deposition of Mn^{2+}. The oxidation peak was observed at applied potential of 1.6 V for all the pH range tested except for pH 2 where the peak shifted slightly to lower potential (1.59 V) while the reduction peak was observed at 0.52 V. The shifting of the current potential was due to the mass transfer of the photon on the surface of the electrode [6]. Other factor might be due to excess of hydrogen ion (H^+) in the solution as low pH value was applied. According to the proposed mechanism suggested by Roushani et al. (2017), H^+ was one of the products during the electrochemical reaction [1].

$$\text{Mn(II) bulk} \longrightarrow \text{Mn(II) absorbed} \longrightarrow \text{Mn(0)} + 2H_2O$$

$$\longrightarrow \text{Mn(II)} + 2H^+ + 2OH^- + 2e^-$$

pH 9 shows the lowest current responses for both the oxidation and reduction peaks. For a clear comparison of oxidation and reduction peak for all range of pH value tested, the current value by which the oxidation and reduction peaks were observed at different pH was plotted as depicted in Figure 1. The oxidation peak was higher in every point of pH value compared to the reduction peak. Among the all pH values, the pH 2 (oxidation peak) showed the most influent by giving the good signal as supporting electrolyte. However, as this research focusing mainly on analytical rather than mechanism of Mn^{2+} electrochemical reaction, thus reduction current response marked by pH 2 was selected as optimal pH condition for further investigation.

IOP Conf. Series: Materials Science and Engineering **606** (2019) 012009 doi:10.1088/1757-899X/606/1/012009

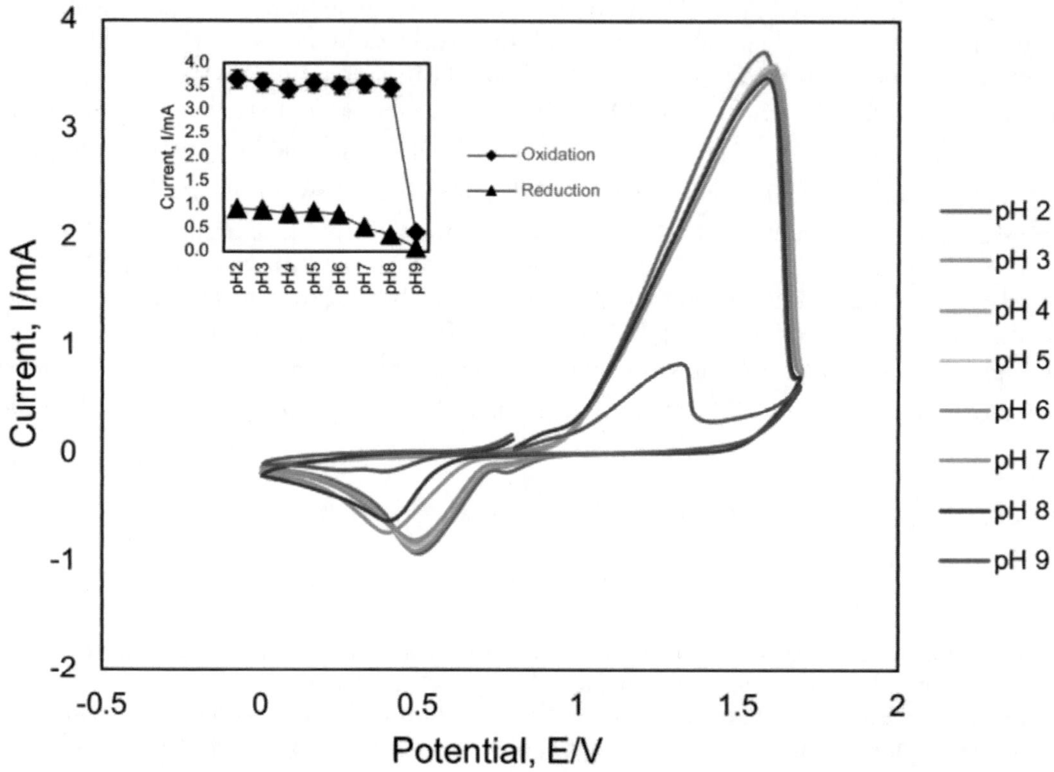

Fig. 1. Cyclic voltammetry test for optimisation of the pH value range from pH 2 to pH 9 for 0.1 M tris-HCl buffer supported by 5 mM Prussian Blue under 100 mVs^{-1}. 0.1 ppm of manganese was directly dilute into the buffer solution and optimised within range 0.0 V to 1.7 V for at least three times experiment was conducted (n>3).

3.2. Effect of different buffers and redox indicators

The current response obtained from the CV measurement of five types of buffer is shown in Figure 2. For this measurement, the buffer concentration and pH were fixed at 0.1 M and 2.0, respectively while 0.5 mM Prussian blue was used as the redox indicator. The buffer provides a complete circuit by acting as electrolytic solution. The volume of the electrolytic solution was fixed at 10 mL with the presence of 0.1 ppm manganese [23]. The selection of manganese concentration was made based on the standard quality level of manganese in drinking water according to the Ministry of Health, Malaysia (2010)[8]. The manganese was diluted directly in the electrolytic solution and tested with bare gold electrode supported by the Prussian blue on the surface of the electrode. Two peaks were formed which located at 1.6 V for oxidation peak while reduction peak was formed at 0.5 V. The current response for oxidation and reduction peaks was 3.439 mA and 0.888 mA, respectively. The sharp of anodic peaks at oxidation range indicates that the process was diffusion-controlled (Filipe and Brett, 2003)[7]. With the highest peak points recorded, 0.1 M tris-HCl buffer (pH 2) was selected as the optimal buffer for further investigation.

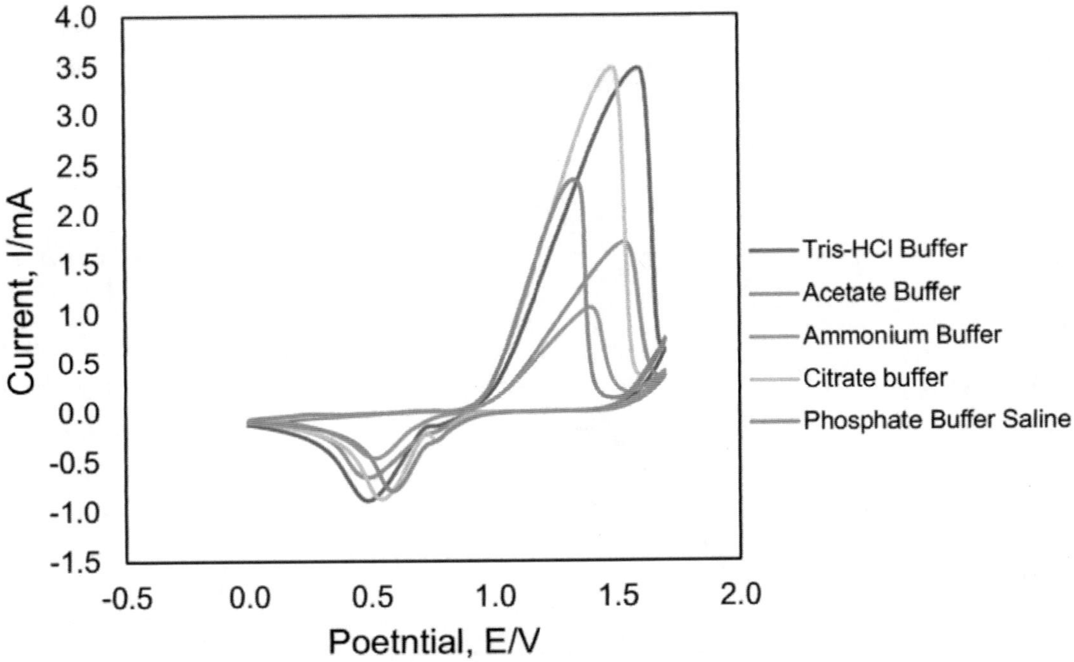

Fig. 2. The cyclic voltammetry test of five types of buffer (0.1 M, pH 2) with 5 mM Prussian blue in the presence of 0.1 ppm in the electrolytic solution. Prussian blue was dropped onto the bare gold electrode under potential range of 0.0 V to 1.7 V. The scan rate was 100 mVs^{-1} while the accumulation time was 5 s. This experiment was repeated at least 3 times for each buffer (n>3).

3.3. Effect of scan rate

Optimisation of the scan rate from 50 mVs^{-1} to 350 mVs^{-1} was conducted to determine the adsorption of the ion in the bare gold electrode. Based on Figure 3, the current response increases with the scan rate. However, for the oxidation peak, unstable current responses were observed as the current movement became faster as the scan rate is increasing. This demonstrated the kinetic of current where the current response is controlled by the diffusion factor [23]. The bell shaped of the current response was changed to a linear line as the scan rate was above 100 mVs-1 on oxidation formation. According to Liu et al. (2011), this indicates that the current response was unstable, which might be due to over-oxidation [27]. On the other hand, for the reduction, stable current responses were observed, and the peak shifted to lower potential value as the scan rate increases. Based on Figure 3, the comparison among the peaks formed at reduction was identified by linear graph. As the increasing scan rate was applied, the peak currents getting higher. The results will influence the next optimisation experiment.

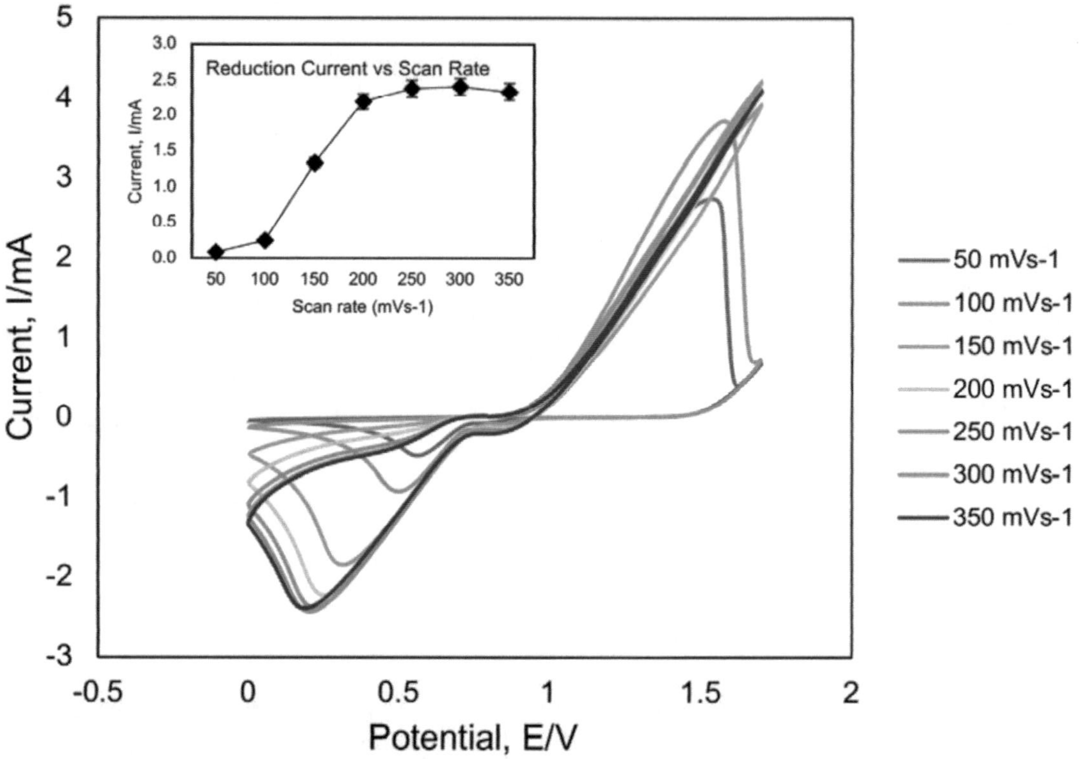

Fig. 3. Cyclic voltammetry test for optimisation of the scan rate range from 50 mVs^{-1} to 350 mVs^{-1} for 0.1 M tris-HCl buffer supported by 5 mM Prussian blue in the presence of 0.1 ppm of manganese ion. Optimisation of the scan rate was conducted at least three times (n>3).

3.4. Effect of accumulation time

Current responses for electrochemical measurement with different accumulation time ranging from 2 s to 40 s were obtained through CV method. The selection of range for accumulation time was made below one minute to fulfil the objective of rapid detection. Accumulation time of 2 s shows the highest current peak which is located at 1.6 V (oxidation range) and 0.46 V (reduction range) as shown in Figure 4. From the same figure, increasing the accumulation time from 2 s to 15 s, gradual decreased of the current peak was observed. As the accumulation time was further increased above 15 s, the current peak increased slightly until 25 s. Then, above 25 s, the current peak was fallen rapidly, and the started to reach a plateau afterward. Based on this observation, stable and high current peak can be obtained in 2 s of accumulation time for oxidation and reduction current responses, suggesting that the method employed in this study capable of detecting the target ion in as fast as 2 s. Therefore, the 2 s was selected for analysis of different concentrations of manganese ion with DPV method. Specifically, reduction peak for 2 s was selected as optimal accumulation time based on the peak responses showed equilibrium state from 2 s until 25 s indicated that Mn^{2+} saturated on the surface of the electrode [1].

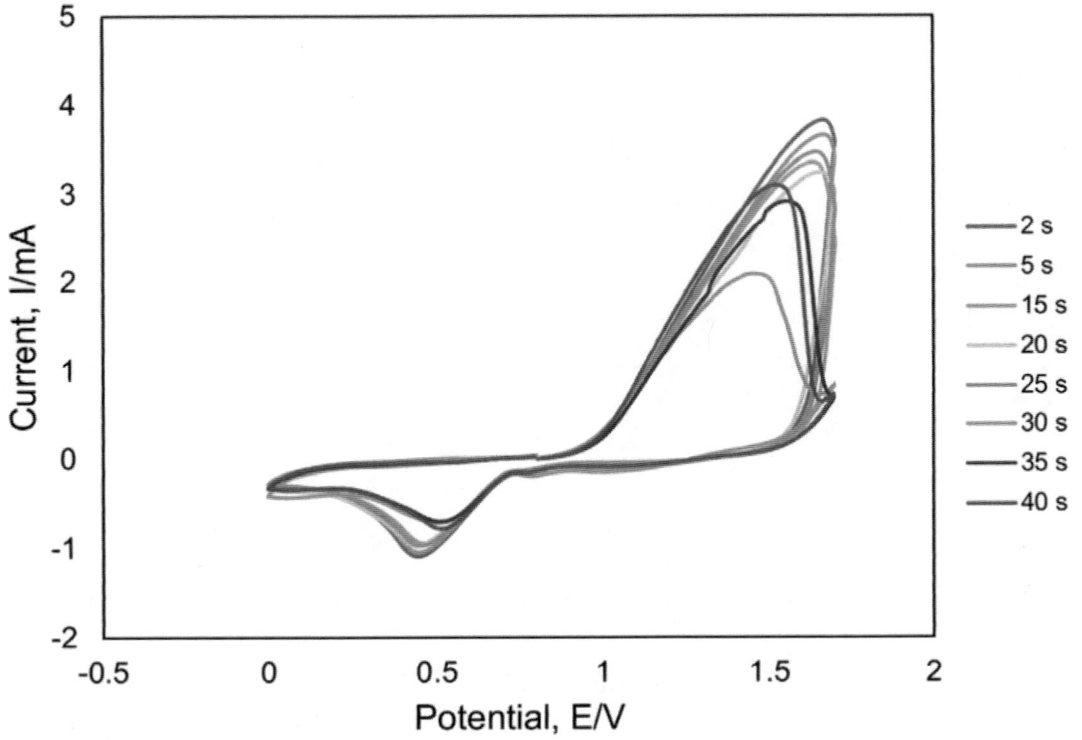

Fig. 4. Cyclic voltammetry test for optimisation of accumulation time from 2 s – 40 s. The experiment was conducted more than 3 times (n>3).

3.5. Determination of different concentrations Mn^{2+} ion

After optimisation step using cyclic voltammetry method, parameters of (0.1 M, pH 2) Tris-HCl buffer, 5 mM Prussian blue, 100 mVs^{-1} scan rate and 2 s accumulation time were found as the optimal condition for detection of manganese ion. These parameters were applied for obtaining the current responses when different concentrations of manganese were added by using DPV method. Figure 5 shows the concentrations of manganese proportional to current responses when the linear of range (LOR) from 0.05 ppm to 1.0 ppm were tested. The linear regression of manganese concentration proportional to current peaks was y = 2.5799x + 26.594 with R^2 = 0.9659 with limit of detection (LOD), S/N=3 was 0.075 mg L^{-1} based on Figure 5(B). The linear regression was selected from every peak point represented different concentrations which showed the line less distributed as the gold electrode supported by Prussian blue. However, the peaks showed positive reduction potential when the peak started to increase as increasing amount of manganese were added [26].

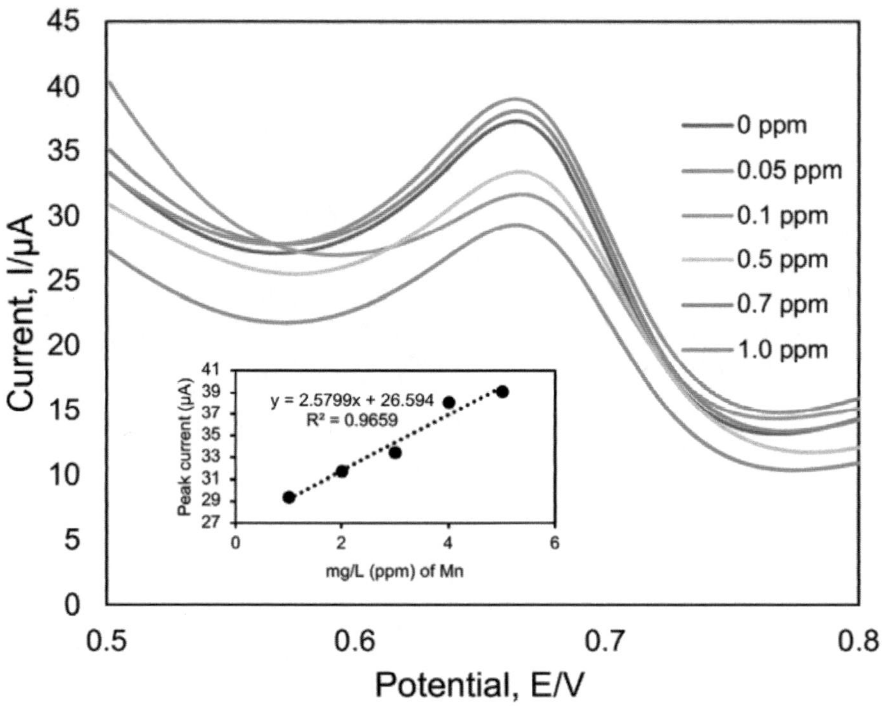

Fig. 5. Differential pulse voltammetry test for different concentrations of manganese from 0.05 ppm to 1.0 ppm in 0.1 M tris-HCl buffer supported by 5 mM Prussian Blue under 100 mVs^{-1}. This experiment was repeated at least three times (n > 3).

4. Conclusion

Electrochemical measurement of manganese ion conducted at the optimized conditions of buffer, redox indicator, pH, scan rate and accumulation time enabled rapid detection of the target ion in 2 s with a satisfactory current response. The use of Prussian blue as redox indicator has increased the current response with shifted peaks of current. The optimization process enabled detection of manganese ion at various concentration when the linear of range (LOR) from 0.05 ppm to 1.0 ppm were tested. The linear regression of manganese concentration proportional to current peaks was $y = 2.5799x + 26.594$ with $R^2 = 0.9659$ with limit of detection (LOD), S/N=3 was 0.075 mg L^{-1}. This research shows the importance of optimization for all the parameters before further modification is applied. The electrochemical method employed in this study can be applied for on-spot detection of dangerous heavy metal ions such as manganese ion in the drinking water.

Acknowledgement

This research was supported by Universiti Malaysia Sabah Research Priority Area Scheme (SBK0248-STWN-2015)

References

[1] Roushani M, Saedi Z, Hamdi F and Dizajdizi B Z 2017 Preparation an Electrochemical Sensor for Detection of Manganese (II) Ions using Glassy Carbon Electrode Modified with Multi-walled Carbon Nanotube Chitosan-ionic Liquid Nanocomposite Decorated with Ion Imprinted Polymer. *Journal of Electroanalytical Chemistry* **804** 1–6
[2] Reazei B, Ghiaci M and Sedaghat M E 2008 A Selective Modified Bentonite-porphyrin Paste

Electrode for Determination of Mn(II) by using Anodic Stripping Voltemmetry. *Sensor and Actuators B* **131** 439-447

[3] Welch C M, Banks C E, Komorsky-Lovri S and Compton R G 2006 Electroanalysis of Trace Manganese via Cathodic Stripping Voltammetry: Exploration of Edge Plane Pyrolytic Graphite Electrodes for Environmental Analysis. *Croatica Chemica Acta* **79 (1)** 27-32.

[4] Wang L, Ding D, Salvi R and Roth J A 2014 Nicotinamide Adenine Dinucleotide Prevents Neuroaxonal Degeneration Induced by Manganese in Cochlear Organotypic Cultures. *NeuroToxicology* **40** 65–74

[5] Alvarez-Bastida C, Martínez-Miranda V, Vázquez-Mejía G, Solache-Ríos M, Fonseca-Montes de Oca G and Trujillo-Flores E 2013 The Corrosive Nature of Manganese in Drinking Water. *Science of the Total Environment* **447** 10–16

[6] Ruggeri S, Terzia F, Zanfrogninia B, Corsia E, Dossib N, Zanardia C, Pigania L and Seebera R 2017 Electroanalytical Determination of Soluble Mn(II) Species at High Concentration Levels. *Electrochimica Acta* **240** 108–113

[7] Filipe O M S and Brett C M A 2003 Cathodic Stripping Voltammetry of Trace Mn(II) at Carbon Film Electrodes. *Talanta* **6** 643-650

[8] Ministry of Health, Malaysia (MoH) 2010 Drinking Water Quality Standard. Drinking Water Quality Surveillance Programme - Ministry of Health, Malaysia

[9] Ravisankar A and Chandrasekaran R 2015 Spatial Distribution of Physico-chemical Properties and Function of Heavy Metals in Soils of Yelagiri Hills, Tamilnadu by Energy Dispersive X-ray Florescence Spectroscopy (EDXRF) with Statistical Approach. *Spectrochimica Acta Part A: Molecular and Biomolecular Spectroscopy* **150** 586–601

[10] Citak D, Tuzen M and Soylak M 2010 Speciation of Mn(II), Mn(VII) and Total Manganese in Water and Food Samples by Coprecipitation–atomic Absorption Spectrometry Combination. *Journal of Hazardous Materials* **173** 773–777

[11] Ferreira S L C, Bezerra M A, Santos A S, dos Santos W N L, Novaes C G, de Oliveira O M C, Oliveira M L and Garcia R L 2018 Atomic Absorption Spectrometry-A Multi Element Technique. *Trends in Analytical Chemistry* **100** 1-6

[12] Jung M Y, Kang J H, Choi Y S, Lee D Y, Lee J Y and Park J S 2019 Analytical Features of Microwave Plasma-atomic Emission Spectrometry (MP-AES) for the Quantitation of Manganese (Mn) in Wild Grape (*Vitis coignetiae*) Red Wines: Comparison with Inductively Coupled Plasma-optical Emission Spectrometry (ICP-OES). *Food Chemistry* **274** 20–25

[13] Gui J, Huang C, Yuan J, Huang Q, Long F, Ma A and Pan J 2018 Handy Cloud Point Extraction Coupled with Inductively Coupled Plasma Mass Spectrometry for Analysis of Trace Zirconium in Complex Biological Samples. *Microchemical Journal* **138** 190–196

[14] Locatelli C and Torsi G 2000 Determination of Se, As, Cu, Pb, Cd, Zn and Mn by Anodic and Cathodic Stripping Voltammetry in Marine Environmental Matrices in the Presence of Reciprocal Interference. Proposal of a New Analytical Procedure. *Microchemical Journal* **65** 293-303

[15] Goodwin A, Lawrence A L, Banks C E, Wantz F, Omanovi D, Komorsky-Lovric S and Compton R G 2005 On-site Monitoring of Trace Levels of Free Manganese in Sea Water via Sonoelectroanalysis using a Boron-doped Diamond Electrode. *Analytica Chimica Acta* **533** 141– 145

[16] Banks C E, Kruusma J, Moore R R, Tomˇc´ık P, Peters J, Davis J, Komorsky-Lovri´c S and Compton R G 2005 Manganese Detection in Marine Sediments: Anodic vs. Cathodic Stripping Voltammetry. *Talanta* **65** 423–429

[17] Jiang T J, Guo Z, Liu J H and Huang X J 2016 Gold Electrode Modified with Ultrathin SnO_2 Nanosheets with High Reactive Exposed Surface for Electrochemical Sensing of As(III). *Electrochimica Acta* **19** 142–148

[18] Giacomino A, Abollino O, Lazzara M, Malandrino M and Mentasti E 2011 Determination of As(III) by Anodic Stripping Voltammetry using a Lateral Gold Electrode: Experimental

Conditions, Electron Transfer and Monitoring of Electrode Surface. *Talanta* **83** 1428–1435

[19] Siddiquee S, Yusof N A, Salleh A B, Tan S G and Bakar F A 2010 Electrochemical DNA Biosensor for the Detection of *Trichoderma harzianum* based on a Gold Electrode Modified with a Composite Membrane Made from an Ionic Liquid ZnO Nanoparticles and Chitosan, and by using Acridine Orange as a Redox Indicator. *Microchimica Acta* **172** 357–363

[20] Zhai X, Li Y, Li J, Yue C and Lei X 2016 Electrochemical Sensor for Detection of Hydrogen Peroxide Modified with Prussian Blue Electrodeposition on Nitrogen, Phosphorus and Sulfur Co-doped Porous Carbons-chitosan. *Materials Science and Engineering: C* **77** 1242-1246

[21] Wen S H, Wang Y, Yuan Y H, Liang R P and Qui J D 2018 Electrochemical Sensor for Arsenite Detection using Graphene Oxide Assisted Generation of Prussian Blue Nanoparticles as Enhanced Signal Label. *Analytica Chimica Acta* **1002** 82-89

[22] Gholivand M B, Akbari A, Faizi M and Jafari F 2017 Introduction of a Simple Sensing Device for Monitoring of Hydrogen Peroxide based on $ZnFe_2O_4$ Nanoparticles/Chitosan Modified Gold Electrode. *Journal of Electroanalytical Chemistry* **796** 17–23

[23] Chaiyo S, Mehmeti E, Zagar K, Siangproh W, Chailapakul O and Kalcher K 2016 Electrochemical Sensors for the Simultaneous Determination of Zinc, Cadmium and Lead using a Nafion/Ionic Liquid/Graphene Composite Modified Screen-printed Carbon Electrode. *Analytica Chimica Acta* **918** 26-34

[24] Gumpu M B, Veerapandian M, Krishnan U M and Rayappan J B B 2017 Simultaneous Electrochemical Detection of Cd(II), Pb(II), As(III) and Hg(II) Ions using Ruthenium(II)-textured Graphene Oxide Nanocomposite. *Talanta* **162** 574-582

[25] Chimezie A B, Hajian R, Yusof N A, Woi P M and Shams N 2017 Fabrication of Reduced Graphene Oxide-magnetic Nanocomposite ($rGOFe_3O_4$) as an Electrochemical Sensor for Trace Determination of As(III) in Water Resources. *Journal of Electroanalytical Chemistry* **796** 33–42

[26] Hwang G H, Han W K, Park J S and Kang S G 2008 Determination of Trace Metals by Anodic Stripping Voltammetry using a Bismuth-modified Carbon Nanotube Electrode. *Talanta* **76** 301–308

[27] Liu P, Zhao H J, Wang L L, Liu Z H, Wei J L, Wang Y Q, Jiang L H, Dong L and Zhang Y F 2011 Analysis of Heavy Metal Sources for Vegetable Soils from Shandong Province, China. *Agriculture Science of China* **10** 109–119

IOP Conf. Series: Materials Science and Engineering **606** (2019) 012010 doi:10.1088/1757-899X/606/1/012010

Spray drying of Mengkudu herbal extract: Effect of wheat flour

K Duduku*, S M Anisuzzaman, B Awang, R Rozianie
Chemical Engineering Program, Universiti Malaysia Sabah, 88400, Kota Kinabalu, Sabah, Malaysia

*E-mail: krishna@ums.edu.my

Abstract. The aim of this study is to determine the antioxidant activity, total phenolic content and total flavonoid content of Mengkudu, and the mixture of Mengkudu with wheat flour. Three major processes are involved in this research such as extraction of Mengkudu, spray drying of Mengkudu extract and the mixture of Mengkudu with wheat flour at 140^{0}C. From results, it was found that the phenolics and flavonoid compounds are present in concentrated Mengkudu extract (CME) and due to these compounds the antioxidant activity of CME was increased. The total phenolic and flavonoid content in the mixture of concentrated Mengkudu extract with wheat flour (MCMEWF) and also in the mixture of spray dried Mengkudu extract with wheat flour (MSDMEWF) are lower than in CME but antioxidant activity of these mixtures was higher compared to antioxidant activity of their individual components. This shows that antioxidant compounds in the mixture exhibit synergistically to donate the hydrogen to DPPH radicals thus increased the antioxidant activity.

1.0 Introduction

Epidemiological studies have shown that regular consumption of whole grains and fruits reduce the risks of various types of chronic diseases such as cardiovascular disease, type 2 diabetes, and cancer [1]. There are various types of phytochemicals found in fruits and grains that help to prevent these kinds of diseases. Phytochemicals are the bioactive, non-nutrient, naturally occurring plant compounds found in fruits, vegetables, and also in whole grains [2,3,4]. The major groups of phytochemicals that may contribute to the total antioxidant capacity (TAC) of plants and foods include polyphenols, carotenoids and traditional antioxidant vitamins such as vitamin C and vitamin E. These phytochemicals may be present in small amounts but are very important to the health of consumers [5]. In the whole grain, the phytochemicals that present consist of phenolic compounds containing one or more aromatic rings and one or more hydroxyl groups carotenoids, and vitamin E, amongst others. Nevertheless, the phenolic content of whole grains had been underestimated, as most of the research only determined the free phenolic content and not the content of phenolics that are bound to cell wall materials [6,7].

Mostly fruits were extracted due to the high antioxidant activities in the fruits. The examples of fruits that are commoly used are sugar beet,cherry,oranges, noni, tamarind, longan, guava and etc. There are also some fruits that can be considered as herbal plant or medicinal plant. These types of plants have medicinal value in the relief and treatment of various aliments [8]. A medicinal plant typically contains a complex mixture of hundreds or more chemical compounds, e.g. alkaloids, glycosides, and flavonoids. For analysis of such a complex mixture, analytical technique with strong separation capability is needed [9].

Due to the useful compounds contain inside the fruits and grain, the researchers have done many fruits and grain extraction to study the properties of these compounds. The correct parameters used in the extraction process definately will give the good result for extraction. The combination of phytochemicals in a fruit can give a significant benefits in health but the combination of phytochemicals from different fruits may give different result. Multiple phytochemicals combinations may result in synergistic activity that increases their bioavailability and their action on multiple molecular targets, thus offering advantages over treatments with single chemical [10]. Synergisms refers to phenomena where the combined effect of independent processes is significantly different from the individual effects considered separately [11]. Many reaserchs were done to determine the synergistic effect of chemical constituents that found in the fruits. The additive and synergistic effects of these biologically active components inside the fruits and grains may be responsible for the health benefits [10]. Objective of this work was to study the spray drying of herbal plant extract on synergistic effect of Mengkudu with wheat flour [13,14].

2.0 Methods

2.1 Preparation of Mengkudu extracts

Mengkudu fruits were washed with tap water followed by washing with distilled water. Then these fruits were cut into small pieces and dried below the sunlight for three days. To remove the moisture that still contained in the dried fruit, it was kept at 65°C in a hot air oven for one day. The extraction of Mengkudu fruits followed the literature procedure with slight modifications [15-17]. Then dried fruit was finely powdered using a mixer. 25 g of powdered sample and 250 ml of ethyl acetate was mixed. This mixture was put in a water shaker bath at 45°C in three days for extraction. After three days, the Mengkudu extract was filtered using filter paper which is Whatman number 1 to separate the supernatant from the residue. The supernatant was put in the oven for one day at 50°C to get concentrated Mengkudu extract. The Mengkudu extract was taken from the oven and stored in a closed container and kept at 5°C in refrigerator before being analysed [18].

2.2 Spray drying of the mixture of Mengkudu extract with wheat flour

100 ml of Mengkudu extract (which equal to 1.5 g) was mixed with 1.5 g of wheat flour that dissolved in 15 ml ethyl acetate to form a mixture. The resulting mixture was stirred continuously before and during the spray drying process to ensure that the samples are fully mixed. A lab scale spray dryer SD-05 with co-current flow was used to spray dried the mixture and in the spray drying process the the inlet gas temperature, feed flow rate and flow rate of drying air were kept constant at 140°C, 250 ml/h and 60 m^3/h respectively. The atomising air is set at a pressure of 1 bar. The same procedure was followed for Mengkude extract with ethlyacetate as solvent.

2.3 Encapsulation yield (EY)

The encapsulation yield was calculated as the ratio of the mass of the microcapsules obtained after spray drying to the mass of the mixture of Mengkudu power with wheat flour [15].

$$EY\% = \frac{\substack{\text{the weight of microparticles} \\ \text{after spray drying}}}{\substack{\text{Total weight of adjuvants and} \\ \text{weight of Mengkudu powder added initially}}} x\ 100 \quad (1)$$

2.4 DPPH radical scavenging activity

DPPH is a stable free radical that reacts with compounds and able to donate hydrgen atom. Thus, the hydrogen donating ability of concentrated Mengkudu fruit extract, the mixture of Mengkudu with wheat flour and spray-dried Mengkudu fruit extract were determined from the change in absorbance at 515nm [19]. To measure the free radical scavenging activity, the samples in methanol solution were prepared by dissolving 10 mg of spray-dried powder in 30 ml of methanol and centrifuging for 10 min using a Sartorius Sigma 3-18 K centrifuge. After that, aliquots of supernatant were added to 3 ml of 0.025 g/l DPPH in methanol. After 38 minutes the change in absorbance was determined by using a 4802 UV- VIS double-beam spectrophotometer. Here, methanol was used as a reference. Besides, DPPH (0–100 mg/l) was used to obtain a standard calibration curve. The percentage of DPPH radical scavenging activity was calculated by using the following equation.

$$\text{DPPH scavenging activity (\%)} = \left[1 - \frac{\text{Abs}_{515}\ \text{sample}}{\text{Abs}_{515}\ \text{DPPH solution}}\right] x\ 100 \quad (2)$$

2.5 Total phenolic content

The total phenolic content was determined with Folin Ciocalteau reagent [17]. Briefly, the samples in methanol solutions were prepared by dissolving 10 mg of spray-dried powder in 30 ml of methanol and then the mixture was centrifuged for 10 min using a Sartorius Sigma 3-18 K centrifuge. Then, 2.5 ml of the 0.2 N Folin Ciocalteau reagent was mixed with the supernatant of the sample extract (0.5 ml) and let it react for 5 minutes. After 5 minutes, 2 ml of 75 g/l sodium carbonate was added to the reaction mixture and diluted to 25 ml using distilled water. Lastly, the reaction mixture was incubated for 2 hours at room temperature and then the absorbance was measured at 760 nm using a 4802 UV-VIS double-beam spec-trophotometer. Here, methanol was used as a reference. Tannic acid (0–100 mg/l) also was used to produce a standard calibration curve. In this work, the total phenolic content was expressed in mg of tannic acid equivalents (TAE/g of spray-dried powder).

2.6.Total flavonoid content

The total flavonoid content was determined using the Dowd method as adopted by Krishnaiah et al. [2]. In order to measure the total flavonoid content, a total of 5 ml of 2% aluminium trichloride (AlC13) in methanol was mixed with the same volume of the extract solution (0.4 mg/ml). After 10 minutes the absorption at 415 nm readings were taken by using a UV-VIS double-beam spectrophotometer against a blank sample which consists of a 5-ml extract solution with 5 ml of methanol without $AlCl_3$. The total flavonoid content can be determined by using catechin (0–100 mg/l) reagent as the standard and the flavonoid content was expressed as mg of catechin equivalents (CE/g of extract).

3.0 Results and Discussion

3.1 Spray-drying of the samples

In this study, two samples were undergone spray drying process such as concentrated Mengkudu extract (CME) and also the mixture of concentrated Mengkudu extract with wheat flour (MCMEWF). During spray drying process the samples were stirred continuously with wheat flour so that the homogenous solution is formed. The inlet temperature of the spray dryer and the feed flowrate were set at 140^{0}C, 318 ml/h respectively.

The powder at the chamber wall were taken by using brush and then the samples were analysed. Table 1 shows that the encapsulation yield (EY) value of spray-dried Mengkudu extract is low which is 2.5% compared to the EY value of the mixture of Mengkudu extract with wheat flour which is 4.5%. The EY values of both samples can be classified as low because both products of the samples were accumulated at the chamber wall due to high moisture content of particles. The wet powders easily can stick at the chamber wall thus affects the EY value.

Table 1. Encapsulation yield of the samples after spray drying

Samples, Spray dried	Weight of micro - particles, g	Weight of adjuvant, g	The weight of wheat flour, g	Mengkudu powder, g	Encapsulatio n yield (%)
Spray dried Mengkudu extract	0.15	3	-	3	2.5
Spray dried Mengkudu + wheat flour	0.27	3	1.5	1.5	4.5

3.2 Antioxidant activity

DPPH assay is simple and inexpensive method that involve the use of 2,2,-diphenyl-1-picrylhydrazyl (DPPH) to measure the atioxidant activity of foods. DPPH is a free radical that is used to test the ability of the compound to act as a free radical scavenger or hydrogen donors and to determine the antioxidant activity of the foods. This method is based on the decrease in absorbance of DPPH free radical at 515nm due to the action of antioxidants. In this study the DPPH scavenging activity of Mengkudu extract, and the mixture of Mengkudu with wheat flour were measured and the results of the assay were expressed in percentage (%) of inhibition of DPPH as shown in Figure 1. From the results we can see that the DPPH radical savenging activity increased in the following order CME>MCMEWF>MSDMEWF>WF>SDME. CME has the highest radical scavenging activity which means that the antioxidant compounds in the sample has highest activity to trap the free radicals or to donate the hydrogen to the free radicals .

CMEWF and SDMEWF have slight difference in percentage of DPPH radical scavenging activity and this is same when compared to the SDMEWF with WF. SDME showed the least DPPH radical scavenging activity which is only 20.05%. By comparing the percentage of DPPH radical scavenging activity of several samples, we can deduce that more antioxidant compounds present in the CME rather than in SDME although both samples are from the same fruit. The decrease in the percentage of DPPH scavenging activity may be due to increase in amount of active components volatilization or deterioration during the spray drying process.

The hydrogen donating activity was higher in CME rather than in WF but when both samples are mixed together, it showed higher DPPH scavenging activity compared to wheat flour. MCMEWF is higher in radical scavenging activity when compared to WF. But when comparing WF with MSDMEWF the difference in radical scavenging activity is not high. MCMEWF and MSDMEWF were higher in radical scavenging activity than WF. Therefore this proves that a large variety of compounds in those mixers act synergistically to trap free radicals.

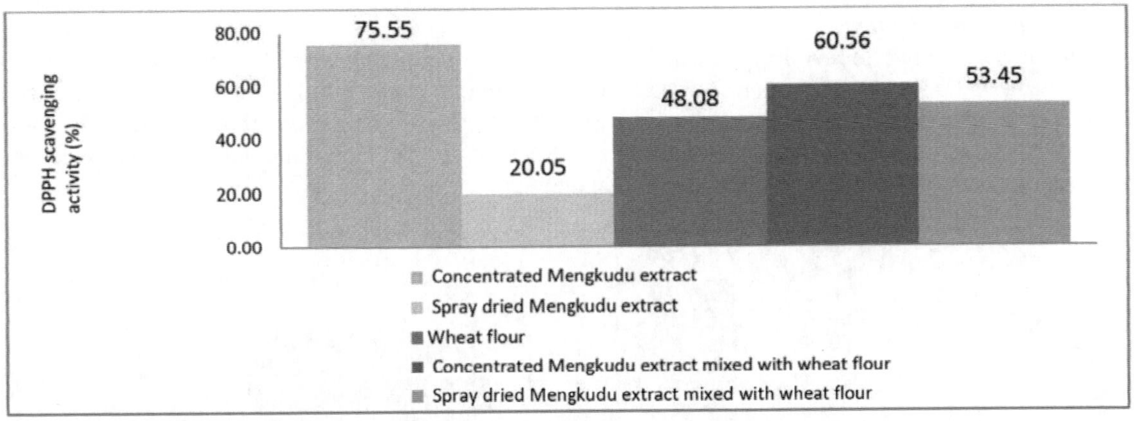

Figure 1. DPPH scavenging activity of Mengkudu extract, wheat flour and the mixtures of Mengkudu with wheat flour

3.3 Total phenolic content

Folin-Ciocalteau method was used in the determination of total phenolic content since many individual phenolic compounds which provide antioxidant activity in fruits cannot be identified and measured by HPLC methods. In this study, the quantative estimation of the phytochemical constituents of CME, SDME, MCMEWF, MSDMEWF and WF were analyzed. Figure 2 shows that the grain product (wheat flour) and medicinal herb (Mengkudu) are rich in total phenols. In this figure CME showed the highest total phenolic content which is 91.67% followed by MCMEWF, MSDMEWF, WF at 67.7%, 45.63%, 38.97% repectively and lastly only 22 mg of TAE/g of SDP of phenolic compounds present in the SDME. There are more phenolic compounds in CME than in SDME. The low content of phenolic compounds in SDME is due to the degradation of antioxidant compounds during drying. The total phenolic content for both CME and WF were compared and it showed that the total phenolic content in CME is greater than the total phenolic content in WF. It is well known that phenolic compounds are highly effective free radical scavenging and antioxidants. The phenolic compounds also rich in MCMEWF and MSDMEWF thus increased the antioxidants activity.

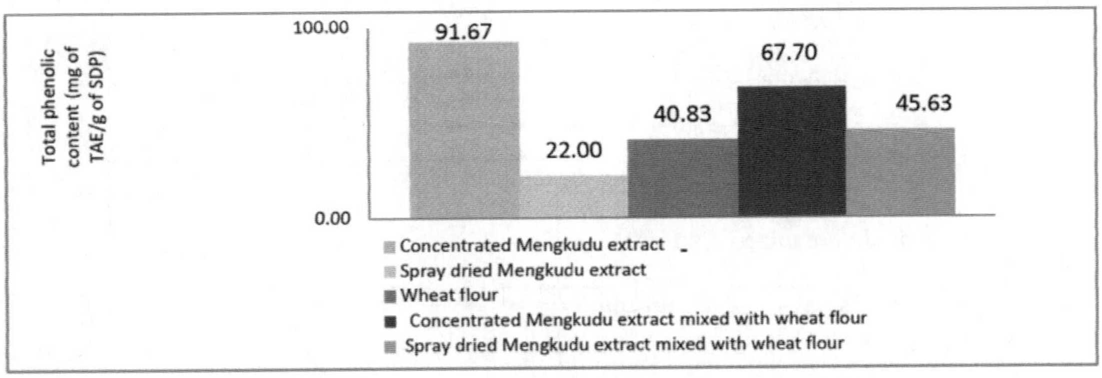

Figure 2. Total phenolic content of Mengkudu extract, wheat flour and the mixture of Mengkudu with wheat flour

3.4 Total flavonoid content

Flavonoids are highly effective free radical scavenging and antioxidants which are commonly used for the prevention and cure of various diseases that are mainly associated with free radicals. The mechanisms of action of the flavonoid compounds are through scavenging or chelating processes [20,21]. In order to quantify the total flavonoids in the samples the Dowd method has been used. In this method Catechin was used as a standard component.

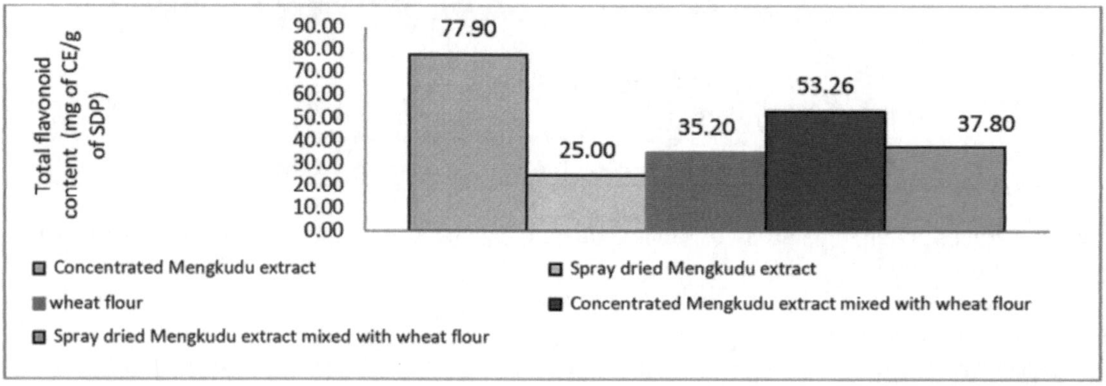

Figure 3. Total flavonoid content of Mengkudu extract, wheat flour and the mixture of Mengkudu with wheat flour

From the Figure 3, the samples that contain flavonoids content are in the following order CMF> MCMEWF> MSDMEWF> WF> SDME. Total flavonoid content was significantly higher in CME and less amount of flavonoid compound in SDME. MCMEWF has higher flavonoid compounds than WF but lower when compared to CME. This is same when we compare MSDMEWF with WF and CME.

3.5 Particle Size

The particle size of the samples were determined by using the particle size analyzer. The samples are dissolved in distilled water before we put into that equipment. From Table 2, it can be seen that the mixture of spray-dried Mengkudu extract mixed with wheat flour has large particle size which is 280.8nm compared to other samples. The smaller particle size of samples is wheat flour which is 113.1nm.

Table 2. Particle size of samples

Samples	Particle size,radius (nm)
Mengkudu powder	138.1
Wheat flour	113.1
The mixture of Mengkudu powder before spray drying mixed with wheat flour	199.3
The mixture of spray-dried Mengkudu extract mixed with wheat flour	280.8

4.0 Conclusion

It can be concluded that more phenolics and flavonoids compound present in CME and due to these compounds the antioxidant activity of CME was increased. The total phenolic and flavonoid content in MCMEWF and MSDMEWF are lower than the CME but antioxidant activity of these mixture were higher compared to their individual components. This showed that antioxidant compounds in the mixture exhibit synergistically to donate the hydrogen to DPPH radicals. In this study, we can deduce that the antioxidant activity increases as the phenolics and flavonoids increase. From the result, the encapsulation yield of the mixture of Mengkudu powder with wheat flour is higher compared to individual components which is 4.5% and the larger particle size of the samples is the particle size of the mixture of spray dried Mengkudu extract with wheat flour which is 280.8 nm.

Acknowledgement
The authors would like to express greatest appreciation to the Centre of Research and Innovation (PPI) for the financial support.

References

[1] Liu C, Xue Y, Ye Y, Yuan F, Liu J and Shuang J 2007 *Agri. Sci. China* **6** 1494-501
[2] Krishnaiah D, Sarbatly R and Nithyanandam R 2011 *Food Bioprod. Process.* **89** 217-33
[3] Krishnaiah D, Devi T, Bono A and Sarbatly R 2009 *J. Med. Plants Res.* **3** 67-72
[4] Rui H L 2004 *J. Nutri.* **134** 3479S–3485S
[5] Jimaima L V, Craige T, Mark W, Naiyana W, Subramanium S and Robert P 2007 *Food Chem.* **101** 1727–41
[6] Adom K K and Liu R H 2002 *Food Chem.* **50** 6182–87
[7] Valentina P, Carmela M, Roberto L S, Gabriele C, Valentino F and Luigi F D 2012 *Food Chem.* **130** 501–9
[8] Vanessa S, Heidi G, Mpho S and Mary G, *Mut.Res.* **42** 35–42
[9] Mohd Z Z, Abdul-Hamid A and Osman A 2001 *Food Chem.* **78** 227–231
[10] Marta G and Arantxa R 2011 *Pharm. Res.* **64** 438– 55
[11] Roberto J B and Behrisch R 1984 *J. Nucl. Mater.* **128 and 129** 764-74
[12] Chong T M, Abdullah M A, Fadzillah N M, Lai O M and Lajis N H 2004 *Plant Cell Rep.* **22** 951–58
[13] Krishnaiah D, Nithyanandam R and Sarbatly R 2014 *Crit. Rev. Food Sci. Nutri.* **54** 449-73
[14] Krishnaiah D, Nithyanandam R and Sarbatly R 2012 *Chem. Eng. Res. Des.* **5** 622-32
[15] Krishnaiah D, Sarbatly R and Nithyanandam R 2011 *J. Appl. Sci.* **11** 2276-83
[16] Krishnaiah D, Bono A, Sarbatly R and Anisuzzaman S M 2015 *J. King Saud Uni. Eng. Sci.* **27** 63-67
[17] Krishnaiah D, Hiaw K B, Sarbatly R, Anisuzzaman S M and Nithayanandam R 2012 *Int. J. Chem. Eng. Appl.* **3** 380- 84
[18] Rene O C, Edemilson C C, Luiza T C, Ezequiane M S, Frederico S, Martins M T, Bara F, Kennia R, Suzana F A and Jose R P 2011 *Food Chem.* **131** 99-105
[19] Blois M S 1958 *Nature* **26** 1199-220
[20] Adriana B, Adriana RC, Chaiana P M, Joseane T, Tatiana M and Andréa A S C 2010 *Brazilian J. Pharma. Sci.* **46** 651-56
[21] Atanassova M, Marinova D and Robarova F 2005 *J. Univ.Chem Technol Metal* **3** 255-60

Preliminary study on oil extraction and biogas production from *Cymbopogon nardus (Serai Wangi)*

E Suali[1], N S I Juasin, F A A Hamit, S M Anisuzzaman and M A Asidin
[1]Faculty of Engineering, Jalan UMS, Universiti Malaysia Sabah, 88400 Kota Kinabalu, Sabah, Malaysia

E-mail: [1]emma.suali@gmail.com

Abstract. An essential oil from *Cymbopogon nardus* has many benefits, specifically in pharmaceutical, cosmetics and perfumery fields. However, extraction of *Cymbopogon nardus* produces abundant waste which usually left to decay naturally. The present study evaluated biogas potential from *Cymbopogon nardus* waste through anaerobic digestion method with rumen fluid as inoculum. The presence of methane which is the main gas that makes up the composition of biogas is verified by using GC-FID and Biogas Analyser. The waste was collected from the oil extraction of various part of *Cymbopogon nardus* plant. The experimental study on the extraction is important to identify the oil yield from various part of *Cymbopogon nardus* plant. The extraction was done through steam distillation with temperature varied from 40 °C to 100 °C. The comparison of oil yield from various part of the plant was done with an aid of GC-MS analysis. It was found that the leaf of *Cymbopogon nardus* contains most essential oil, which gives the highest oil yield about 1.5 % at 100 °C followed by the stem (0.3 %). The oil yield found in the plant flower was less than 0.3 % and none found in the root. Extraction at 100 °C resulted in higher oil yield compared to 40 °C, 60 °C and 80 °C. The citronellal content was the highest compound presents in the oil followed by geraniol. The average generation of biogas on the other hand were range from 0.4 ml/day to 12.5 ml/day. The F/I ratio which produced the highest volume of biogas is 10 (w/w).
Keywords: Citronella; *Cymbopogon nardus*; Essential oil; Biogas; Extraction; Steam distillation

1. Introduction

Cymbopogon nardus or known as *Serai Wangi* in local dialect of Malaysia and Indonesia is non-edible grass-like plant but has many benefits and uses [1]. In Malaysia, *Cymbopogon nardus* was mostly used as natural insect repellent, natural insecticides and household products such as detergent, soap and so on [2]. In recent studies, essential oil of *Cymbopogon nardus* were found to contain antibiotics for aquaculture [3-4]. Since antibiotic was banned for aquaculture, the oil from *Cymbopogon nardus* could be used as an alternative to commercial antibiotic.

Citronella oil contains many elements such as citronellal (16.9 %), citronellol (10.4 %), elemol (9.1 %) and geraniol (3 %) [5], whereas citronellal gives the distinctive lemon scents while geraniol give the rose-like odor and citrus smell. Citronellal which is the main content of oil yield is widely used for fragrance and perfumeries [1]. It was also reported useful for medicinal purposes. The benefits of the *Cymbopogon nardus* can be optimized through extraction of its essential oil. Extraction of *Cymbopogon nardus* is the process to segregate substances from its sources in order to obtain the essential elements. The extraction process can be performed either in small or large scale under controlled temperature and pressure [6].

There are many methods to extract essential oil. This include supercritical fluid extraction (up to 2.8%) [9] steam or hydro distillation (less than 1%) [7,8] and solvent-less extraction (up to 1.3%) [10]. The oil yield from the plant is affected by the extraction method and the parameters used such as temperature, pressure, duration for extraction, and types of solvent. Hence, in order to achieve a high yield of oil, suitable method must be chosen and the parameter must be set in optimum conditions.

The empty husk of the extracted *Cymbopogon nardus* usually left to decay naturally as well as the others part which not included in extraction process. It was also used as natural fertilizer for its own cultivation. This produces abundant waste to obtain a minute amount of essential oil from its leave. For every metric ton of citronella leaves distilled, only 8 kg of essential oil is produced leaving 992 kg of biomass residue discarded as a waste [11]. Thus, it is useful if ones can find a way to reuse the waste. In this study, a waste from *Cymbopogon nardus* extraction was tested for it potential as biogas producer. Biogas is a form of renewable energy that can be harvested from organic substrates which has acquired a lot of attention in the research field for decades [12]. Biogas generation through anaerobic digestion reduces emission of greenhouse gases compared to the burning of fossil fuels [13]. The process of anaerobic digestion is one of the oldest method to convert organic substance into biogas. Hence, anaerobic digestion aids a process of waste treatment and stabilization [14]. Biogas produced from organic substrate depends on the content, chemical composition and degree of biodegradability of the substrate used [15]. The performance of anaerobic digestion depends on few conditions namely hydraulic retention time, temperature, carbon to nitrogen (C/N) ratio, organic loading rate, partial pressure, pH, nature of the organic substrate, inoculum ratio and presence of oxygen in anaerobic environment [16].

The potential of various biomass as inoculum has been studied and considered for biogas production [17]. Different inoculum behaves differently in terms of the ability of converting organic substrate into biogas. It was reported that optimal feed to inoculum (F/I) ratio is within 1 to 10 [18]. Thus, in this study, a residue of *Cymbopogon nardus* extraction which include leaves, stem, root and flowers are tested for methane production. The residue include leaves, stem, root and flower of *Cymbopogon nardus*. It was claimed by many that no oil in the flower of *Cymbopogon nardus*. However, no scientific report to support the claim. Thus, it is interesting to conduct an oil extraction on the *Cymbopogon nardus* flower, stem and root besides leaves. In this study, oil yield of various parts (leave, stem, root and flower) of *Cymbopogon nardus* was compared.

2. Materials and Methods

2.1 Preparation of Cymbopogon nardus

Cymbopogon nardus was taken from Ladang Serai Tuaran at Kampong Selupoh, Tuaran, Sabah. The roots, stems, flowers and leaves of fresh *Cymbopogon nardus* plant was cut into a smaller pieces and weighed before drying 48 hr at 60 °C in the oven. The sample is finely grounded and stored in a sealed plastic bag and left in a cabinet at a room temperature without the presence of sunlight to prevent any possible oxidation reaction. The extracted husk was prepared according to the method reported in Alfa et al. [19].

2.2 Preparation of rumen fluid

Rumen fluid was used as inoculum. It was obtained from the Sabah Meat Technology Centre, Kinarut, Sabah. The rumen was used to provide sufficient bacteria for the digestion process. The method to prepare the inoculum was adapted from Ahmed and Kazda [20]. Inoculum that was freshly taken shortly after the process of slaughtering was sieved in order to remove the large particles and kept at 37 °C under anaerobic conditions for several days for degassing purpose.

2.3 Extraction of Cymbopogon nardus

The extraction of *Cymbopogon nardus* was conducted to determine the oil yield and oil content.

The extraction was performed by using steam distillation method as mentioned in [7-8]. The experiment was conducted in a Clevenger type distillation apparatus using water and hexane as solvents. The solvent-solvent extraction was performed using separatory funnel and the removal of hexane from the mixture of oil - hexane was performed using rotary evaporator. About 1.5 L of distilled water was added into the 2 L pear-shaped glass contained 150 g of sample. The straight glass condenser was connected to the pear-shaped glass. The temperature was set from 40 °C to 100 °C for the extraction of stems, flower, leaves and roots.

2.4 Anaerobic digestion

Potential of *Cymbopogon nardus* waste as biogas feedstock was studied and the experiment was performed in a batch operation. One hundred grams of waste feedstock was fed into respective biodigester and mixed with 100 ml water on a 1:1 w/v basis. The inoculum was fed into the digester according to the respective feedstock to inoculum ratio (F/I). The sample was kept in the bottle which acts as a mini bioreactor and left at a room temperature. The bioreactor A1 and A2, B1 and B2, C1 and C2, D1 and D2, and E1 and E2 were prepared with F/I ratio of 2, 2.5, 3.3, 5 and 10, respectively. The performance of the digestion was measured by the cumulative production of biogas. The study was conducted in duplicate in order to minimize the error and to avoid the risk of not getting results due to leakage considering samples in gaseous form is hard to contain and control. The solid retention time (SRT) of the anaerobic digestion was 40 days followed the method conducted by Owamah et al. [21] and poultry droppings. Measurement was taken at an interval of 2 days for each digester at 4 pm for a duration of 40 days.

2.5 Analytical procedures

2.5.1 Gas Chromatography Flame Ionization Detector (GC-FID)

The gas type analyses were performed using GC-FID, Agilent Technologies GC 7890B with HP 5MS column. Initial temperature of 105 °C and detection temperature of 150 °C. Gas samples were tested using manual injection by using Agilent manual syringe with the volume of 10 ml.

2.5.2 Biogas analyzer

The composition of the biogas was confirm using Geotech's Biogas 5000 Portable Biogas Analyzer. The device is reliable in measuring gas levels of methane, carbon dioxide, oxygen and hydrogen sulfide.

2.5.3 Gas Chromatography-Mass Spectrometry (GC-MS)

The oil content analysis was performed to detect the presence of valuable component which are citronellal, citronellol and geraniol. Citronella oil obtained from the experiment was analyzed using Agilent 6890/5973 GC-MS, Model G1530A. The analysis was performed by analyzing the peaks obtained from the GC-MS which represent the composition of citronella oil. The composition include citronellal, citronellol and geraniol. The presence of the valuable components are detected by comparing its retention time with the commercial standards solution with a brand of Eau de Senteur. The GC-MS composed of the capillary column which has 0.25 internal diameter, 30 m long, 0.25 μm film thickness. The gas carrier is helium with the flow rate of 1.0 mL/min and the temperature is set up varied from 50 °C to 300 °C at 8 °C/min with the injector temperature of 280 °C. The oven temperature was at 80°C.

3. Results and discussions

3.1 Essential oil yield of Cymbopogon nardus

The yield of *Cymbopogon nardus* oil ranged from 0 % to 1.5 % whereas oil yield found in leaves was the highest about 1.5 % followed by the stem approximately 0.28 %. It was found that there are oil yield obtained from the flower about 0.24 % and none from the root. The oil in aromatic plant is stored in a special glands called glandular gland. These special gland located in different parts of plant but mostly found in the leaves of the plants. Hence, more oil is obtained from the leaves compared to the other parts. Other than that, higher yield of oil observed from leaves may be due to less fibers in leaves which can store more oil compared to stem.

The ages of the plant affect the oil color which the younger plant produces lighter color of oil compared to the old aged plant. The old plants are richer in more resinous and have darker oils due to the continuing evaporation of the lighter fraction of the oil. Hence, more components are presence in the oil from leaves and stem compared to the flowers. However, the characteristics of the oil obtained from the flowers similar to the oil obtained from the leaves and stem. It only varied in term of color whereas the flowers produces light-yellow oil compared to oil obtained from the leaves and stem which have dark-yellow color.

3.2 *Effect of temperature on extraction yield*
The extraction temperature is one of the most important variable in the extraction process as its affect the amount of product yield. Thus, in this study, effect of temperature was conducted on the oil yield of the leaves. The leaves was selected than the others part of the *Cymbopogon nardus* because it has the highest oil yield. It was found that the highest extraction temperature was at 100 °C, which gives the highest percentage oil yield about 1.53 %, followed by the extraction temperature of 80 °C and 60 °C which gives the percentage oil yield of 0.6 % and 0.54 %, respectively. The lowest amount of oil yield was obtained at 40 °C, which only 0.03 %. Table 1 presents the comparison oil yield of the leaves at various temperature.

Table 1. Oil yield with respects to extraction temperature

Temperature (°C)	Mass of oil (g)	Oil yield (%)
40	0.050	0.03 ± 1
	0.051	
	0.049	
60	0.543	0.543 ± 1
	0.541	
	0.542	
80	0.905	0.603 ± 1
	0.903	
	0.902	
100	2.295	1.530 ± 1
	2.294	
	2.293	

It is showed that the temperature affect the oil yield and higher extraction was obtained at higher temperature compared to the lower temperature. This is due to viscosity of the oil which get lowered at high temperature causes more oil released from the intact cell.

3.3 GC-MS analysis of essential oil components
The concentration for each constituent is different for each sample. There are also others compounds presents beside essential oil. Others GC-MS analysis from stem and flowers were not shown. The presence of others peak besides essential oil compounds showed that the oil extracted contains other impurities. However, the impurities were not identified for this study. It would be interesting to study in the future about others potential of *Cymbopogon nardus* extract to produce others compound besides citronellal, citronellol and geraniol. The summary of peak appearance represents essential oil components for this study is shown in Table 2. Table 2 shows concentration of compound presence in extracted citronella for each component extracted from various parts of the plant. The presence of citronellal, citronellol and geraniol compounds in the oil extracted were determined by comparing with

the retention time and the spectra fragmentation of the standards, which appeared around 13.8 min,14.8 min and 15.6 min for citronellal, citronellol and geraniol, respectively. The concentration of the compounds were obtained by substituting the area under the curve (y) of the compound into the linear equation obtained from the calibration curve for the respective compound. From the study, all compounds have high concentration in leaves followed by stem and flowers. The empty husk from the extraction was further treated for biogas production.

Table 2. Concentration of compounds in various part of the plant

	Compounds type		
Plant Components	Citronellal	Citronellal	Geraniol
Leaves	10.36	12.31	13.56
Stems	9.54	10.51	11.32
Flowers	6.26	5.78	7.68

3.4 Effect of F/I ratio on biogas yield

There are 10 bioreactors labelled as A1 to E1 and A2 to E2 were prepared to study the potential of waste husk *Cymbopogon nardus* to generate biogas. The daily gas generated from the bioreactors were summarized in Figure 1. The volume of biogas were quantified by measuring the volume of water displaced. In the duration of 40 days, approximately a total of 309 ml of biogas was produced from various reactors. Volume of biogas produced differs from each reactor. As can be seen, the reactors A1, A2, B1, B2, D2, and E2 shows no displacement of water. Despite no displacement of water, no conclusion can be made yet prior to analyzing the gas in the headspace of the reactor using GC-FID and Biogas Analyzer. The aforementioned reactors may produce biogas as well but the volume is not significant, hence no water displacement. Although, the fact that the reactors may have leakage can also be considered. The reactors which showed a significant water displacement were C1, C2, D1, and E1. The average production of biogas from each biogas-producing reactors are 0.6 ml/day, 0.4 ml/day, 1.95 ml/day and 12.5 ml/day from the reactor C1, C2, D1, and E1, respectively.

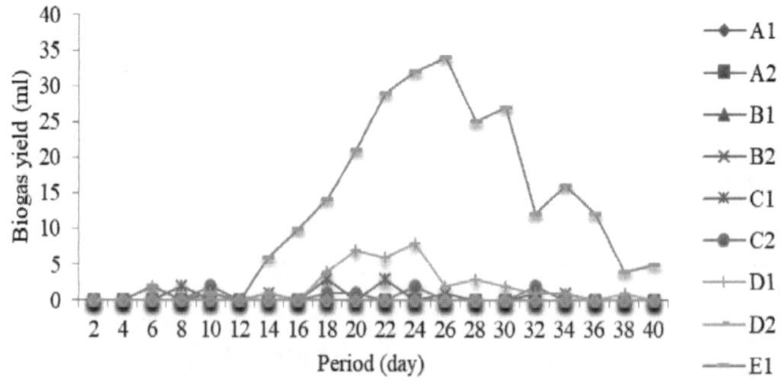

Figure 1. Daily biogas production from bioreactors

In general the production of biogas started around day 6 and showed gradual increase until day 26. From day 0 until day 6, there is no gas produced. The late start in gas production could be attributed to the nature of the substrate where presence of lignin is the limiting factor of the anaerobic digestion process. This situation similar as reported in Alfa et al. [19] where the substrate for lignin degradation prior to digestion increases decomposition rate of substrate. Substrate such as citronella grass containing lignin decomposes at a slower rate compared to animal intestinal wastes which have undergone prior digestion within the digestive system of animals. Reactor E1 with F/I of 10.0 showed the most significant volume of gas produced among all of the reactors. Reactor E1 generated the highest cumulative volume of gas with 250 ml. As the F/I ratio increases, the biogas generated also increases. Low efficiency of gas production might be due to insufficient methanogens. Optimal F/I ratio must be reached in order to accelerate the methanogen activities and start-up of biogas activity. The F/I of 10.0 generated the most amount of biogas while the decreased in F/I also showed decreased in biogas production.

Inoculum reacts differently with different substrate due to the differences in chemical compositions and/or physical structures. The findings in the current study however cannot be compared to any previous study related to inoculum ratio variation on *Cymbopogon nardus* for biogas production as there is a lack of data and no published data on the notion. From the current study, the F/I ratio which produced the highest volume of biogas is 10.0. Study varying the range up to a higher F/I ratio should be further investigated to find the optimal F/I ratio as F/I ratio below 10.0 showed insignificant amount of gas produced.

3.5 Validation of methane gas in biogas sample using GC-FID and biogas analyzer

The validation of methane gas was conducted for each reactor. There are two methods were applied to analyze the contents of gas presents in each bioreactor, GC-FID and biogas analyzer. GC-FID is the most preferable method as it accurately shows the type of gas in sampling. However, GC-FID does not detect gases that present in non-standard gas (air). Thus, further testing was conducted using biogas analyzer. The first analysis using GC-FID shows that only reactor A1 and D1 contains methane. The analysis result under GC-FID is shown in Figure 2. Figure 2 shows the peak appearance of methane gas for both A1 and D1. There are no peak for the others sample which were taken from each bioreactor. This probably due to the gas present which was not of standard gas (air), thus cannot be detected by the FID detector. Based on GC-FID analysis, it can be confirmed that the reactors; A1 and D1 contained methane gas. As the highest peak can be found around 2.315 to 2.865 minutes which confirms the presence of methane according to the standard reference. The A1 samples showed a delayed in the highest peak appearance which was at around 2.865 minutes due to technical difficulties while conducting manual injection using the GC-FID. To support the gas analysis result from the GC-FID, further testing was conducted by using biogas analyzer. Biogas analyzer is used to measure the fractions of gases presents in the sample. The analysis results of biogas analyzer were summarized in Table 3 with focuses on methane, carbon dioxide, oxygen and hydrogen sulfide.

A1

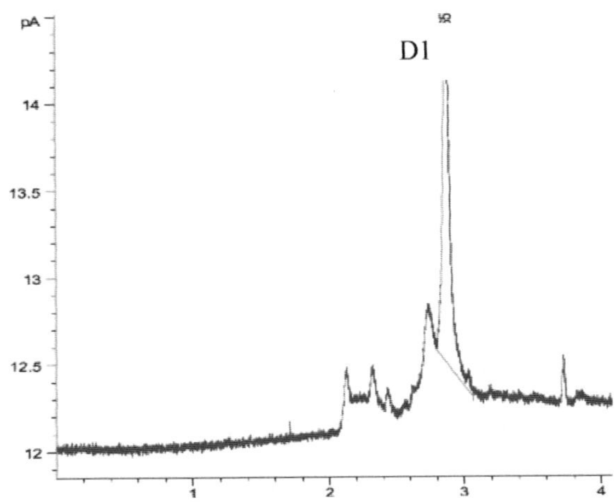

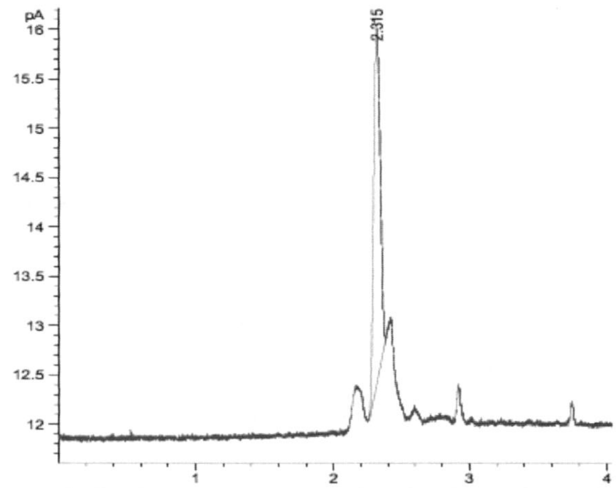

Figure 2. Peaks appearance under the GC-FID analysis

The reactor with the highest CH4 content is from the reactor D1 at 10.1 % while the lowest is from C2 at 5.9%. The highest content of CO_2 produced is from B1 with a percentage of 38.1 % and the lowest from A1 with 11.0%. The reactor containing the highest oxygen is from reactor A1 with a percentage of 15.8 % while the lowest is from D1 with 0.7 %. Impurity gas which is H_2S is also present in the gas where reactor E1 contained the highest with 52 ppm while the lowest one is 4 ppm from reactor A1. When the ratio of F/I is increased, it is predicted that the H_2S content will also increases due to increasing sulfur content. However, the formation of hydrogen sulfide in the gas samples does not follow a specific increasing trend when in theory, as the F/I is increased, the H_2S content should follow as well. It is important to determine the physicochemical properties of the substrates prior to the study in order to predict the H_2S content in the gas sample.

Table 3. Biogas composition in each bioreactor

Reactor	Biogas composition			
	CH_4 (%)	CO_2 (%)	O_2 (%)	H_2S (ppm)
A1	7.8	11.0	15.8	4
A2	7.5	26.6	3.8	18
B1	8.1	38.1	11.6	24
B2	8.7	27.4	3.5	12
C1	6.5	26.4	2.3	14
C2	5.9	27.3	2.0	8
D1	9.6	31.0	0.7	19
D2	10.1	19.6	6.8	11
E1	6.3	22.5	4.5	52
E2	7.8	26.4	1.2	21

The typical constituent of biogas is methane (50 % to 80 %), carbon dioxide (25 % to 50 %), nitrogen (0 % to 10 %), hydrogen (0 % to 1 %), hydrogen sulphide (0 % to 3 %) and oxygen (0 % to 2 %). The result obtained using the analyzer does not correlates with the literature. There is no general trend as well in terms of the composition analyzed based on the variation of F/I. Therefore, the effect of F/I variation on the biogas composition is inconclusive. Errors in the experiment is probably attributed to the nature of the sample, where it is hard to detect contamination and leakage. Leakage of sample to the atmosphere may even contaminate and affect the accuracy of the result obtained.

4. Conclusions

It was proven that, the *Cymbopogon nardus* leaves, stem and flower contains oil. However, the highest oil yield was from the leaves. Citronellal was the major compound of the extracted oil. The waste of *Cymbopogon nardus* was also confirm to produce biogas especially methane.

Acknowledgement

This work was supported by the UMS Scheme Grant SGI0082-2018.

References

[1] Chen W and Viljoen A M 2010 *S. Afr. J. Bot.* **76** 643
[2] Kpoviessi S, Bero J, Agbani P, Gbaguidi F, Kpadonou-Kpoviessi B, Sinsin B and Quetin Leclercq J 2014 *J. Ethnopharmacol.* **151** 652
[3] De Toledo L G, Ramos M A, Spósito L, Castilho E M, Pavan FR, Lopes É, Zocolo G J, Silva F A, Soares T H, Dos Santos, A G, Bauab TM and De Almeida M T 2016 *Int. J. Mol. Sci.* **17** 1252
[4] Wei LS and Wee W 2013 *Iran J. Microbiol.* **5** 147
[5] Kamari FEL, Taroq A, Atki YE, Aouam I, Oumokhtar B, Lyoussi B, Abdellaoui A 2018 *Int. J. Pharm. Sci. Rev. Res.* **50** 14
[6] Ruenroengklin N, Zhong J, Duan X, Yang B, Li J and Jiang Y 2008 *Int. J. Mol. Sci.* **9** 1333
[7] Hawthorne SB, Grabanski CB, Martin E and Miller DJ 2000 *J. Chromatogr. A* **892** 421
[8] Khajeh M, Yamini Y and Shariati S 2010 *Food Bioprod. Process.* **88** 227
[9] Manaf M A, Mustapa AN and Mustapa K 2013 IEEE Business Engineering and Industrial Applications Colloquium (BEIAC), 7-9 April, Langkawi, pp. 73-78
[10] Spigno G, Tramelli L and De Faveri DM 2007 *J. Food Eng.* **81** 200
[11] Manurung R, Melinda R, Abduh MY, Widiana A, Sugoro I and Suheryadi D 2015 *Pak. J. Nut.* **14** 919
[12] Achinas S, Achinas V and Euverink GJW 2017 *Eng.* **3** 299
[13] Scarlat N, Dallemand JF and Fahl F 2018 *Renew. Energ.* **129** 457
[14] Ariunbaatar J, Panico A, Esposito G, Pirozzi F and Lens PN 2014 *Appl. Energ.* **123** 143
[15] Raposo F, De la Rubia MA, Fernández-Cegrí V and Borja R 2012 *Renew. Sust. Energ. Rev.* **16** 861
[16] Mao C, Feng Y, Wang X and Ren G 2015 *Renew. Sust. Energ. Rev* **45** 540
[17] Raposo F, Banks CJ, Siegert I, Heaven S and Borja R 2006 *Process Biochem.* **41** 1444
[18] Tumutegyereize P, Muranga FI, Kawongolo J and Nabugoomu F 2011 *Afr. J. Biotechnol.* **10** 18243
[19] Alfa IM, Okuofu CA, Adie DB, Dahunsi SO, Oranusi SU and Idowu SA 2012 *Int. J. Green Chemist. Bioprocess.* **2** 34
[20] Ahmed S and Kazda M 2017 *Anaerobe.* **46** 114
[21] Owamah HI, Alfa IM and Dahunsi SO 2014 *Renew. Energ.* **68** 366

IOP Conf. Series: Materials Science and Engineering **606** (2019) 012012 doi:10.1088/1757-899X/606/1/012012

Investigating the potential of meltblown polypropylene nanofiber membrane for desalination by membrane distillation

C K Chiam*, R Sarbatly, S Widyaparamitha
Membrane Technology Research Group, Material and Mineral Research Unit, Faculty of Engineering, *Universiti Malaysia Sabah*, Jalan UMS 88400 Kota Kinabalu, Sabah, Malaysia

*E-mail: chiamchelken@ums.edu.my

Abstract. A tubular nanofiber membrane module is developed and tested in the seawater desalination by using air-gap membrane desalination (AGMD). Meltblowing device is used to produce the nanofibers which are collected at a distance of 37 cm away from the tip of the spinneret. Results show that the flux varies almost constantly with the feed flow rates. The independence of the flux on the feed flow rates suggested that the meltblown nanofiber membrane is a dense membrane in which the volume of pores has been reduced and compacted significantly. The flux increases exponentially with the feed temperatures which obeys the Arrhenius relation. The quality of permeate is essentially improved by using the compacted nanofiber membrane. The rejection of the dissolved salts is more than 99.3%, in which the conductivity values of permeate ranged between 8 and 50 μS/cm.

1. Introduction

Membrane distillation (MD) has been recognized as a novel desalination technology which is competitive with the commercial reverse osmosis process. MD is attractive recently because it has met the requirement of process intensification such as small footprint, simple flow sheet as less mechanical parts are used, non-pressurized process, feasible to be operated with renewable energies and waste heat, consuming less energy which leads to lower the cost [1-2].

Generally, MD is categorized into four types; they are direct contact MD, air-gap MD, sweeping-gas MD, and vacuum MD. In the desalination process by using MD technique, the seawater is usually heated to the temperatures around 50 – 90°C [3-5]. The heated seawater is contacted directly with the membrane surface on the upstream side. The membrane used in MD is porous and hydrophobic; thus only the water vapor is allowed to transport through the membrane pores. The temperature on the downstream side of the membrane surface is maintained lower than the upstream side; hence the water vapor pressure difference across the membrane drives the water vapor transport through the membrane. The water vapor is then condensed and collected as pure water. Although few MD pilot plants have been developed under projects MEDINA, MEDESOL and MEDIRAS funded by European Commission; membranes especially for MD application are still unavailable commercially.

Nanofiber membrane is newly received attention for MD application in recent years [6-7]. Nanofiber membrane is attractive for MD because of its high porosity due to the interconnected open pore structure, high permeability thus resulting in high flux, and ease for tailoring the membrane thickness. By meltblowing technique, polypropylene (PP) pellets are melted at high temperature to produce the molten and the molten is sprayed through a spinneret to form nanofibers. Nanofiber membrane is formed when a sheet comprises of the nanofibers overlap with each other randomly.

Meltblown nanofiber membrane has been successfully tested for AGMD by our research team lately [8]. One of the significant findings from our previous study is that the 'loose' structure of the nanofiber membrane did not permit the AGMD to operate at high feed flow rates because severe membrane leaking took place easily. However, operating at low feed flow rate and low feed temperature resulted in low fluxes but allow the relatively high permeation of dissolved salts through the nanofiber membrane which eventually reducing the salt rejection. Similarly, increasing the nanofiber membrane thickness (0.55 - 0.75 mm) has definitely reduced the flux but the corresponding permeate conductivity value increased. The conductivity values by using the 'loose' structure of nanofiber membrane ranged between 200 and 1520 μS/cm [8].

One of the methods to reduce the 'loose' effect in the meltblown nanofibers is to collect the nanofibers immediately once they are sprayed out from the tip of the spinneret. The meltblown nanofibers at the nearer to the tip of the spinneret are hotter and softer. Thus, the membrane formed by these nanofibers can have more 'compacted' structure which is possible can reduce the membrane wetting and improve the salt rejection. The current study aims to fabricate the meltblown tubular PP nanofiber membrane at a shorter distance between the spinneret tip and the collector, than the previous study [8].

2. Experimental methods

2.1. Materials

The polypropylene (PP) pellets used to fabricate nanofiber membrane were supplied by Sun-Allomer, Japan. The pellets were used without any further treatment. A polypropylene mesh was used as support of the nanofiber membrane. The isopropyl alcohol (IPA) supplied by QRëc was used for characterizing the nanofiber membrane porosity.

2.2. Fabrication of nanofiber membrane by meltblowing

The meltblowing device (Japan Zetta Co. Ltd.) as illustrated in Figure 1 was used in this study. The PP pellets were poured into the feeder; the pellets were extruded and melted at three different stages of temperatures: 250, 280 and 300°C. The molten was sprayed through a spinneret with diameter of 0.8 mm. A rotating pipe was located at 37 cm away from the tip of the spinneret. A tubular nanofiber membrane was formed with the inner diameter and length were 2.1 cm and 96 cm, respectively. The tubular nanofiber membrane was wrapped with the PP mesh support, as shown in Figure 2.

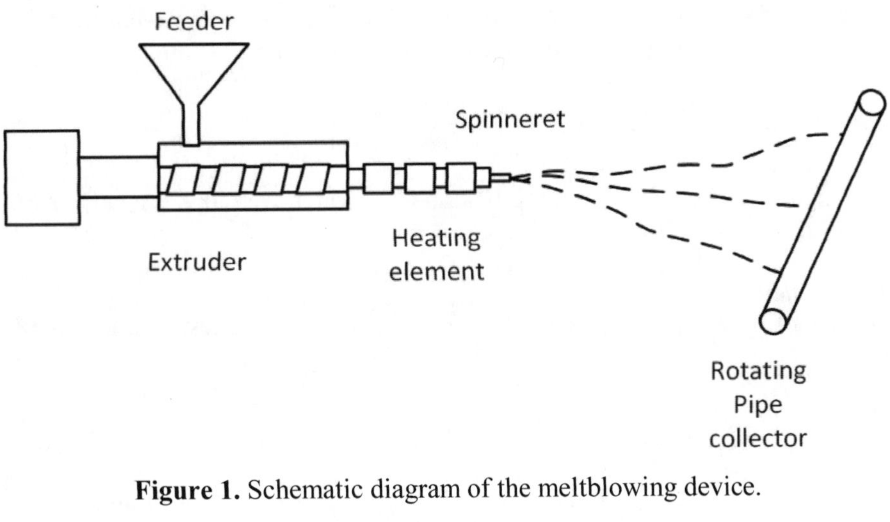

Figure 1. Schematic diagram of the meltblowing device.

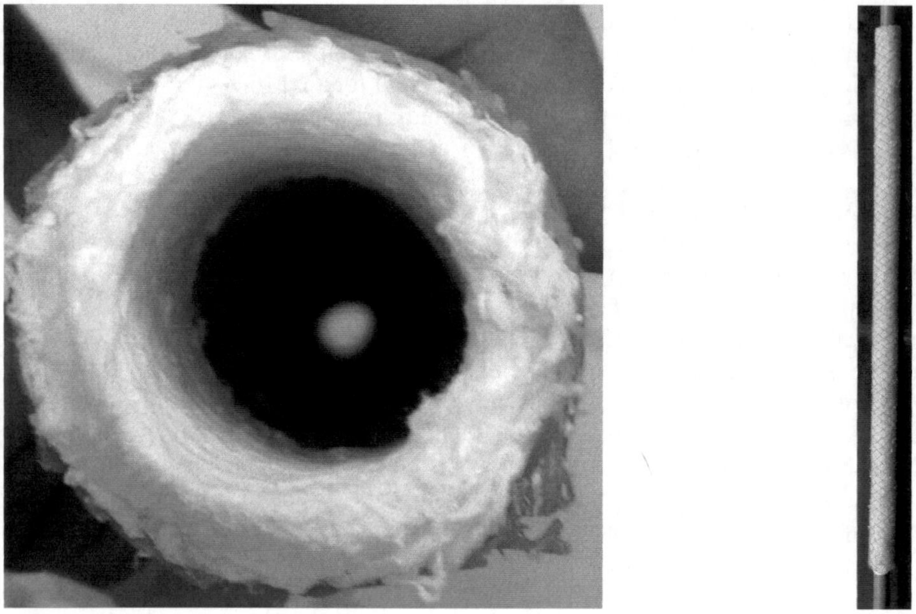

(a) (b)

Figure 2. (a) Cross view of the meltblown PP tubular nanofiber membrane. (b) Side view of the meltblown PP tubular nanofiber membrane.

2.3. Characterization of membranes

The morphology and the size of the nanofibers were investigated by a scanning electron microscope (SEM, Hitachi).

The thickness of the nanofiber membrane was determined by using a digital micrometer (MDC-100M, Mitutoyo, Kawasaki, Japan), with a precision of ±0.001 mm, and the mean value of 20 locations together with its standard deviation were computed.

The membrane porosity is defined as the volume of the pores divided by the total volume of the membrane. The membrane porosity was computed by the gravimetric technique in this work. A small sample of nanofiber membrane in dry condition was cut and its weight was recorded as W_{d}. The same nanofiber membrane sample was then immersed in the IPA for 1 min to allow all of the voids were fully filled up. The weight of the IPA-wetted nanofiber membrane was recorded as W_{w}. Next, the membrane porosity was determined by using the equation as follows:

$$\varepsilon = 100 \times \frac{(W_{\mathrm{w}}-W_{\mathrm{d}})/\rho_{\mathrm{IPA}}}{(W_{\mathrm{w}}-W_{\mathrm{d}})/\rho_{\mathrm{IPA}}+W_{\mathrm{d}}/\rho_{\mathrm{PP}}} \quad (1)$$

where ρ_{IPA} and ρ_{PP} were the densities of the IPA and PP, which were 0.786 g/cm3 and 0.946 g/cm3, respectively.

2.4. AGMD experiment

The meltblown nanofiber membrane with effective area 0.063 m2 was tested in the AGMD system as shown in Figure 3.

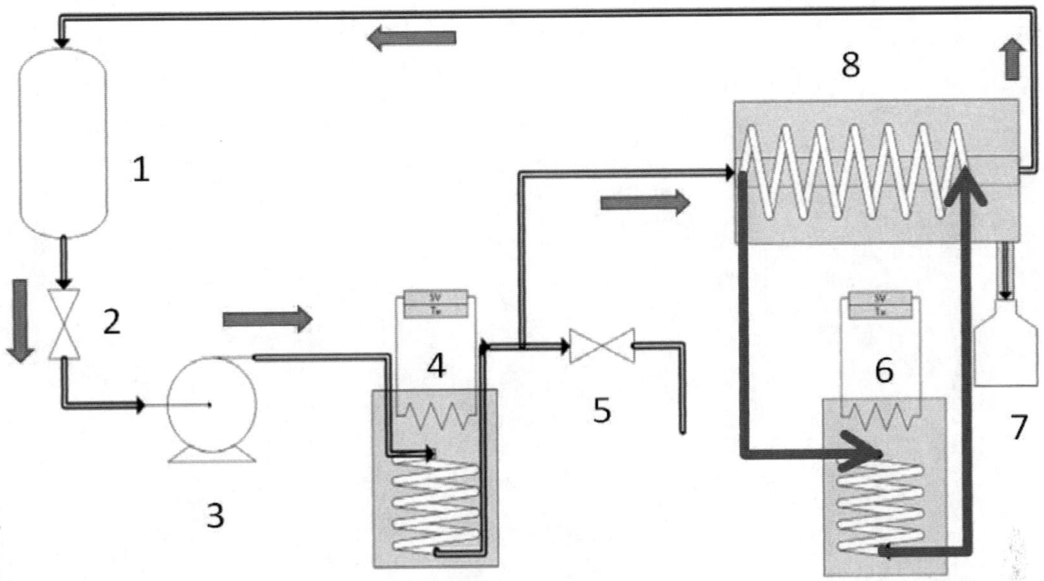

Figure 3. Schematic diagram of the AGMD experiment setup. 1, feed reservoir; 2, valve; 3, pump; 4, temperature control bath with heating coil; 5, sampling valve; 6, temperature control bath with cooling coil; 7, permeate collector; 8, nanofiber membrane module.

Seawater was sampled from a beach near the Universiti Malaysia Sabah. Two liters of the seawater were poured into the feed reservoir. The seawater was circulated through the lumen of the tubular membrane by a gear pump completed with a digital flow meter (Micro pump Gear Pump Drives 75211-35, Cole-Parmer Instrument Cp., Illinios, USA) and passing through a water heating bath (Model 9012A12E, Polyscience, Niles, IL USA). Cooling water was circulated through a copper coil and the coil wound along the tubular membrane module without touching the membrane surface. A chiller circulation system (Thermo Haake, DC 30) was used to maintain the cooling water temperature at 15°C. The tested feed flow rates were 350, 400, 450, 500 and 550 mL/min; while the feed temperatures were 40, 50, 60, 70 and 80 °C. The experimental flux was computed by using the equation as follows:

$$J = \frac{m_{\mathrm{p}}}{A \cdot t} \qquad (2)$$

where m_{p} was the weight of permeate, A was the membrane area and t was the time for collecting the permeate.

3. Results and discussion

3.1. Effect of the distance collector on the fiber size

Based on the SEM image shown in Figure 4, the size of the most of the meltblown fibers ranged from 0.3 to 4 μm which falls in the same range of the order of magnitude when the distance collector away from the tip of the spinneret is 240 cm [8]. It is suggested that the distance between the tip of the spinneret and the collector has no essential effect on the size of nanofibers.

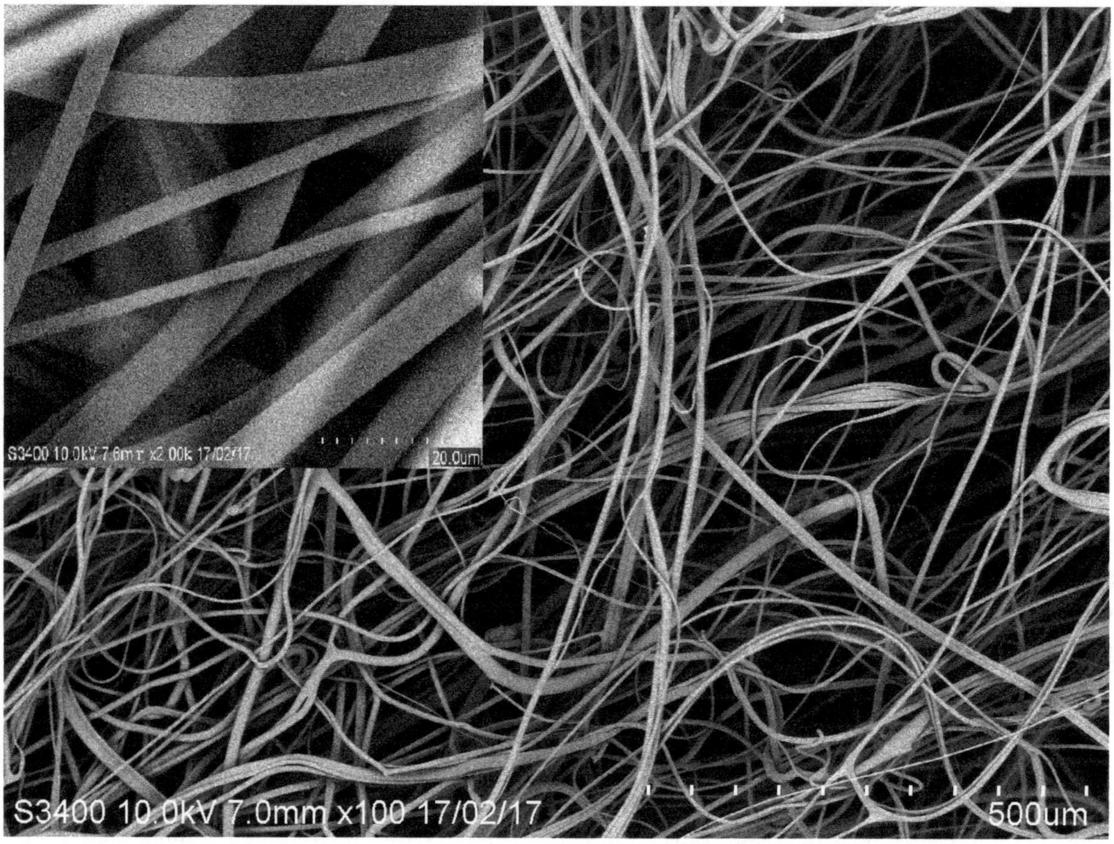

Figure 4. SEM images of the meltblown PP nanofiber membrane.

3.2. Effect of operating conditions on AGMD flux

The effect of the feed flow rate on the AGMD flux at feed temperature 70°C for 2 h operation is shown in Figure 5. The fluxes are nearly not influenced by the feed flow rates ranging from 350 to 550 mL/min. In a comparison with our previous study [8], the independence of the flux on the feed flow rate designates that the structure of the nanofiber membrane used in this study has become denser in which the similar observations have been reported in various MD systems [2].

The effect of the feed temperature on the AGMD flux at feed flow rate 350 mL/min for 2 h operation is presented in Figure 6. The results show that the flux increases with increasing the temperature of the feed seawater. The flux increases exponentially with the feed temperature which is fitted to the Arrhenius type dependence, with the correlation coefficient approximately 0.97. This pattern of relationship between the flux and feed temperature is commonly found in other MD system [9-11].

3.3. Permeate quality

In this study, the denser meltblow nanofiber membrane formation shows that the volume of pores of the nanofiber membrane has been significantly reduced and compacted when the distance collector is reduced to 37 cm. The average porosity of the nanofiber membrane is approximately 86%, which is smaller than those collected at about 240 cm away from the spinneret tip [8].

The membrane resistance is the essential parameter to affect the flux because the membrane thickness contributes a considerable resistance to the water vapor transport across the dense membrane. The thickness of the meltblown nanofiber membrane used in this work is 3.2 ± 0.6 mm. Attractively, this nanofiber membrane has successfully improved the quality of permeate. The conductivity values of condensed permeates varied between 8 and 50 μS/cm while the conductivity value of seawater is 7338 μS/cm which resulting in the rejections of dissolved salts are at least 99.3%.

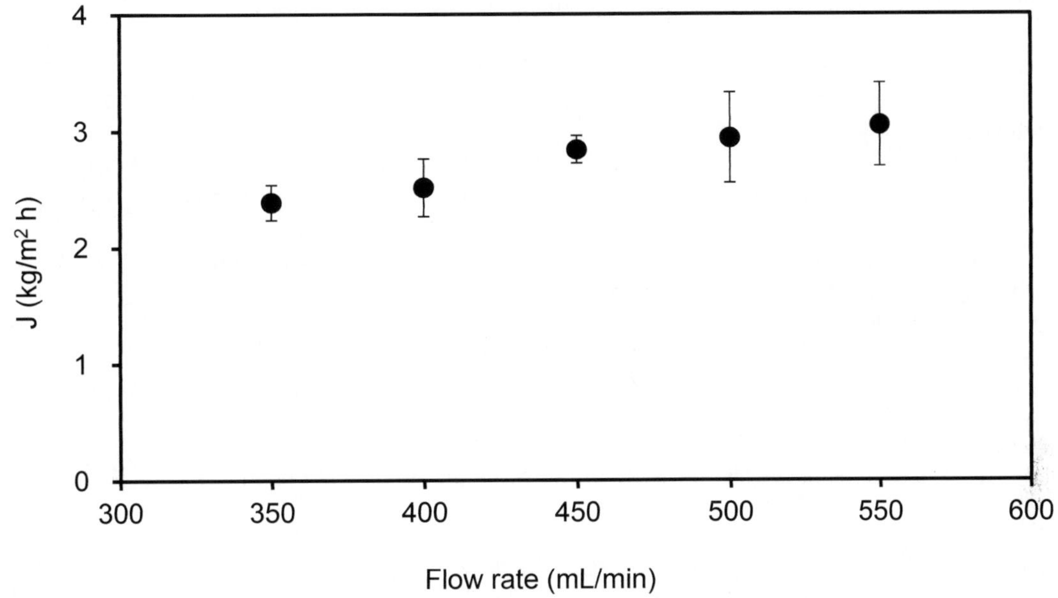

Figure 5. The effect of the feed flow rate on the AGMD flux. (Feed temperature: 70°C; water cooling temperature: 15°C; Operation duration: 2 hours)

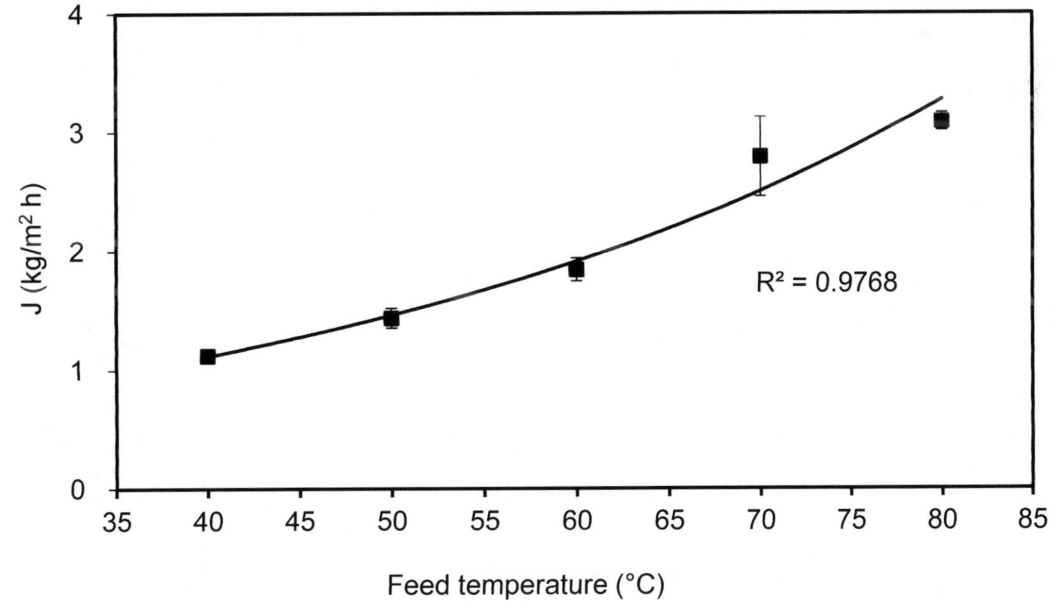

Figure 6. The effect of the feed temperature on the AGMD flux. The solid line is the Arrhenius fit of the data. (Feed flow rate: 350 mL/min; water cooling temperature: 15°C; Operation duration: 2 hours)

4. Conclusions

This work evaluated the improvement of permeate quality in an AGMD seawater desalination by using the meltblown tubular polypropylene (PP) nanofiber membrane which the nanofibers are formed and collected at a distance of 37 cm away from the tip of spinneret. The results show that:

(1) The diameter of the nanofibers ranged from 0.3 to 4 µm.
(2) The volume of pores is reduced to approximately 86% as compared with the nanofiber membrane formed at 240 cm away from the tip of the spinneret (which is approximately 90%).
(3) The effect of the feed flow rate (350 - 550 mL/min) on the AGMD flux is negligible. It is suggested that the membrane is dense which is due to the volume of membrane pores reduced and compacted.
(4) The AGMD flux increases exponentially with increasing the feed temperature which fits the Arrhenius type dependence.
(5) The membrane thickness is 3.2 ± 0.6 mm.
(6) The rejection of the dissolved salts is above 99.3% in which the conductivity values of condensed permeate ranged between 8 and 50 µS/cm.

Acknowledgements

The authors also wish to thank the laboratory facilities provided by the Universiti Malaysia Sabah.

References

[1] Alkhudiri A, Darwis N and Hilal N 2012 *Desalination* **287** 2
[2] Chiam C K and Sarbatly R 2013 *Chem. Eng. Process.* **74** 27
[3] Zheng R, Chen Y. Wang J, Song J, Li X M and He T 2018 *J. Membr. Sci.* **555** 197
[4] Chen X, Gao X, Fu K, Qiu M, Xiong F, Ding D, Cui Z, Wang Z, Fan Y and Drioli E 2018 *Desalination* **443** 212
[5] Khalifa A E and Alawad S M 2018 *Desalination* **437** 175
[6] Khayet M, García-Payo M C, García-Fernández L and Contreras-Martínez J 2018 *Desalination* **426** 174
[7] Park S H, Kim J H, Moon S J, Drioli E and Lee Y M 2018 *J. Membr. Sci.* **550** 545
[8] Sarbatly R, Chiam C K, Widyaparamitha S, Chang Y W and Lee C A 2016 *IOP Conf. Ser. Earth Environ. Sci.* **36** 1
[9] Chiam C K and Sarbatly R 2014 *Chem. Eng. Process.* **79** 23
[10] Chiam C K and Sarbatly R 2016 *Chem. Eng. Process.* **102** 169
[11] Mengual J I, Khayet M and Godino M P 2004 *Int. J. Heat Mass Transf.* **47** 865

IOP Conf. Series: Materials Science and Engineering **606** (2019) 012013 doi:10.1088/1757-899X/606/1/012013

Analysis of Evaporation Data from an Enclosed Pond Ventilated by Solar Chimney

***R Md. Mizanur*[1], *R Md. Ashikur*[2], *M L Noor Ajian*[3], *Chi Ming Chu*[2]**
[1]Mechatronics Engineering Department, World University of Bangladesh
[2]Chemical Engineering Programme, Faculty of Engineering, Universiti Malaysia Sabah
[3]Mechanical Engineering Programme, Faculty of Engineering, Universiti Malaysia Sabah

Email: chrischu@ums.edu.my

Abstract. Water removal from industrial effluent streams, salt cultivation farms, biodiesel conversion process from microalgae, and food drying by using pond evaporation with a higher rate of mass flux can be considered as an economical and energy saving scheme. The objective of this study is to utilize a method of enhancing the pond natural evaporation rate to improve the sustainability of industrial operations. In this study, canopied ventilation technology was employed to enhance the pond evaporation. This technology caused the natural draft enhancement by installing a wire mesh on the chimney. The unit consists of a solar irradiated solid wall chimney installed on the evaporation pan as a means to increase the evaporation rate of existing water body. Experiments were conducted in square pan at the same size as class-A pan in three configurations. The experimental result showed that the evaporation flux from configuration YY (Solar chimney with wire mesh) averaged at 13 per cent more than configuration NN (Open Pond) and at 20 per cent more than configuration YN(Solar chimney without wire mesh). Counter to intuition, for configuration YN the evaporation averaged at 6 per cent lower than the evaporation from the open pond configuration. The pan evaporation flux was found to be predicted satisfactorily by the Rayner[5] evaporation model showing it is able to aggregate reasonably the effect of changing characteristics of the three configurations in terms of net radiation, vapor pressure deficit and natural ventilation draft.

1. Introduction

Evaporation is one of the important processes in many fields ranging from hydrology and agriculture, to food science and engineering applications. Methods of enhancing the practice of natural evaporation in industrial operations employing the process will improve sustainability. The evaporation rate can be increased through a number of mechanical and chemical means, but notably raising the water temperature, increasing the exposed surface area (Ahmed et al.[1]). Jhajharia et al.[2] examined the influence of different meteorological parameters on pan evaporation at Agartala, Bangladesh and using the linear and exponential methods concluded that the wind speed and mean temperature have a positive influence on pan evaporation. Any obstructions, near the pan, may alter the wind structure over the pan site and increase deposition of foreign matter (Pereira[3]). Roderick et al.[4] and Rayner[5] suggested that the changes in wind speed were the most important cause for the decline of pan evaporation rates. Liu et al.[6] concluded that air temperature variation dominated the change in pan evaporation, which offset the effect of wind speed and led to the increase in the pan evaporation.

This research investigates an enclosed pond evaporation by solar irradiated ventilation through an enhanced chimney in East Malaysia. The augmentation involves the incorporation of a solar chimney ventilation technique using wire mesh developed by Chu et al[7] in a test facility consisting of a solar irradiated ventilated class-A pan (SINVAP).

2. Factors influencing pond natural evaporation

Chow et al. [8] stated that the mechanism of transporting the vapor from the water surface plays a great role in the phenomenon of evaporation from open water body. To estimate evaporation properly, meteorological variables are required, such as net radiation, wind speed, relative humidity and air temperature have been widely used (Shuttleworth [9]). According to Liu and Xia [10], the rate of evaporation is most responsive to net radiation, followed by relative humidity, air temperature and wind speed. Smith and Lavis [11] indicated that water temperature is one of the most important factors for evaporation. The existing vapor pressure gradient between the water surface and air above is considered the main driving force in the evaporation process (Jackman and Yotsukura [12]). Vapor pressure deficit (VPD) is the difference between the amount of moisture in the air and how much moisture the air can hold when it is saturated.

The earliest water surface evaporation model is that of Penman [13], consisting of vapour pressure deficit and wind speed as parameters. Thom et al's [14] pan evaporation model added a wind function for when the wind speeds were only in the range of 0 to 3 m/s. Rayner [5], based on the field measurements in Australia, then proposed another wind function for higher wind speed.

3. Experimental Facility

The evaporation process took place in a Class-A pan as specified by the National Weather Service, U.S.A., modified with a square instead of a round shape. The pan evaporation data were completed for three configurations during 26th May to 7th July, 2015 at UMS. A square shaped pan was used to observe the water evaporation at 6.0367°N; 116.1186°E. This latitude locates Universiti Malaysia Sabah (UMS). The observation had been carried out daily 8:00am to 5:00pm. The summary of the experimental pan configuration is shown in Table 1.

Table 1: Summary of the experimental configurations

Conf.	**Pan structure**			**Solar Chimney Ventilation**	**Wire mesh**
	G.I. sheet	**Length (mm)**	**Height (mm)**		
NN	0.7mm	600	254	No	No
YN	0.7mm	600	254	Yes	No
YY	0.7mm	600	254	Yes	Yes

NN = No Chimney and no Wire Mesh, Open Pan
YN = With Chimney but no Wire Mesh
YY = With Chimney and with Wire Mesh

The experimental pans were installed on a 150 mm height wooden platform set on the ground in a grassy location at the Faculty of Engineering in Universiti Malaysia Sabah to reduce the heat exchange between the bottom wall of the pan and the soil. The photograph of the configuration NN is shown in Figure 1.

A solar chimney was used to ventilate vapor from the enclosed pan of configuration YN and configuration YY (Figure 1). A square pan was placed inside an enclosure. The topside of the enclosure was covered with a transparent plastic sheet to transmit solar energy of the radiation to the pan water. The parallel sides of the enclosure were covered by the polymer sheet to make it prevent air ventilation to the surrounding. An inlet opening of 10.5-inch diameter was installed at the front end of the enclosure to supply fresh air from the ambient into the enclosed pan. A duct was used to convey the moist air from the enclosed pan into the solar irradiated chimney as air rose through the latter by natural convection. The schematic diagram and photograph of the configuration YN and configuration YY are shown in Figure 1 and Figure 2 respectively.

IOP Conf. Series: Materials Science and Engineering **606** (2019) 012013 doi:10.1088/1757-899X/606/1/012013

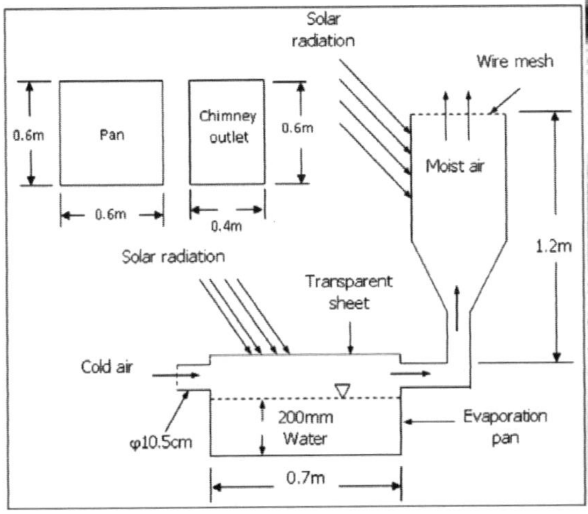

Figure 1: Schamatic view of configuration YN (No me **Figure 2**: Schamatic view of configuration YN (No mesh) and

Figure 2: Photograph of the experimental set up for configuration YN (Left) and configuration YY (Right)

4. Experimental procedure

The pan was filled with fresh water to the datum of the pan, where the water height in the pan was 200 mm for each run of all configurations. Referring to Figure 1, the ambient cold air entered in the enclosed pan by the combined effect of wind and buoyancy pressure through the inlet opening. When the ambient air passed over the pan water, the air would be heated by solar irradiation bringing the vapor to a higher temperature. Hereafter the moist air was ready to enter the duct junction that was connecting the enclosed pan to the solar chimney. The heated moist air would rise up in the direction of the outlet opening of the chimney by natural convection. The chimney wall was made of hollow polymer sheets and could therefore insulate effectively the warm moist air rising from the evaporation pan. Due to the buoyancy effect of moist air in the chimney there was a negative pressure difference across the working height between the moist air outlet and the entry, with the consequence of the flow direction towards the chimney outlet opening.

A Kestrel 4000 pocket Weather & Environment meter, a vane anemometer (Airflow LCA 30 VA), a CMP3 Pyranometer, an Infrared thermometer, K-type thermocouples, an eight channel Cole Parmer data logger, a multi-meter and a digital depth gauge were the principal instruments. A Kestrel 4000 pocket Weather & Environment meter was installed at the 65cm height above the ground surface to measure the local wind speed over the evaporation pan. The Kestrel 4000 Weather & Environment meter also was used to collect local air temperature, local relative humidity, local atmospheric pressure and local elevation. The vane anemometer was installed at the inlet duct of the enclosed pan to measure the average air flow through the enclosed pan. A CMP3 Pyranometer was used to measure the net solar energy incidence to the pan water. The change of water level in the pan through evaporation was measured by a digital depth gauge. The infrared thermocouple was used to measure the water surface temperature. Two sets of combination of four thermocouples were located at the outlet of chimney to measure the average outlet air temperature covering approximately equal segments of the cross-sectional flow area.

Table 2: Instruments range and accuracy

Instrument	Parameter	Range	Resolution	Accuracy
Kestrel 4000	Wind speed	0.6 to 60 m/s	0.1 m/s	3% of reading
	Ambient temperature	-10 to 55 °C	0.1 °C	0.5 °C
	Relative humidity	0 to 100%	0.1%RH	3% RH
	Pressure	10 to 1654.7 mbar	0.1mbar	1 mbar
Anemometer (LCA30 VA)	Flow velocity	0.25 to 30 m/s	0.01 m/s	1% of reading
K-type thermocouple	Temperature	-200 to 1250 °C	0.01 °C	0.75% above 0 °C & 2% below 0 °C
Depth gauge	Water level	0 t0 450 mm	0.01 mm	Counting value composition error
Infrared thermometer	Water temperature	-32 to 380 °C	0.1 °C	5 % below 0 °C & 2% above 0 °C
CMP3 Pyranometer	Solar radiation	Max. irradiance 2000 W/m²		10 % of daily sums

5. Results and Discussion

The prevailing ambient conditions comprising of net radiation, air temperature, wind speed, relative humidity and atmospheric pressure are typical of an equatorial climate where the net radiation ranged from 7.96 to 435 Wm^{-2}, air temperature from 23.6 to 37.7°C, atmospheric pressure from 100.45 kPa to 101.17 kPa, wind speed from 0.71 ms^{-1} to 1.31 ms^{-1} and relative humidity (RH) from 50.8% to 95.2 %. The vapour pressure deficit and maximum air temperature values were always in the descending order of YY, YN and NN. Evaporation rate however, had a descending order of YY, NN and YN.

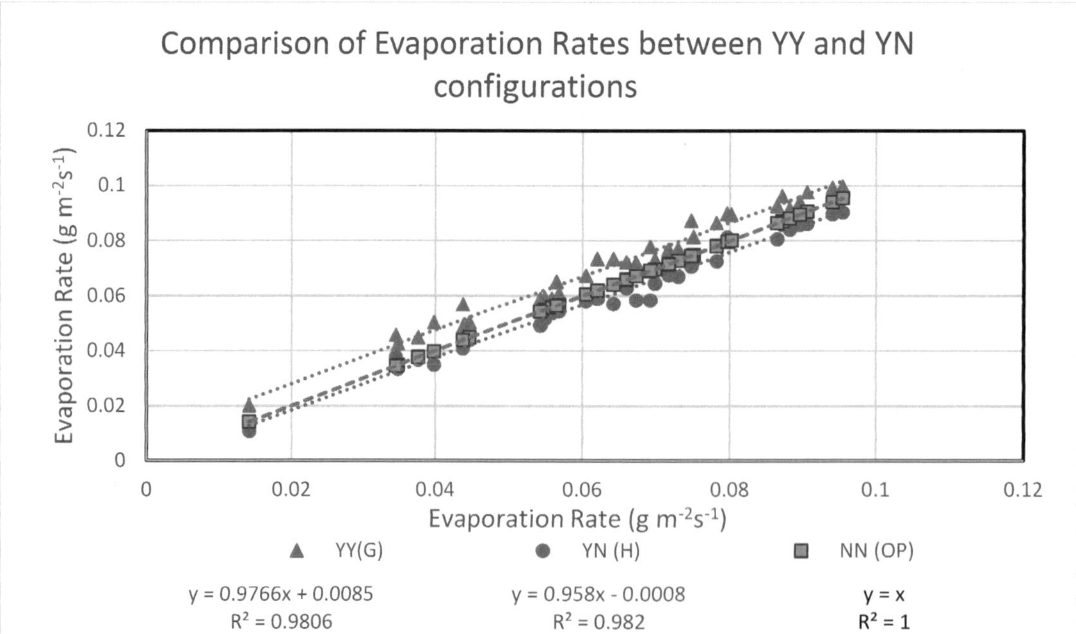

Figure 3: Relationship of experimental evaporation flux of the configurations

The net radiation in the enclosed pan was slightly below that of the net radiation in the open ambient pan water although the trend were closely mimicking each other. The cause for the difference in net radiation values between the open pan and the YY and YN configurations was the YY and YN material radiation transmissivity. The evaporation pan in the open atmosphere received direct solar radiation; in contrast, the evaporation pan in the YY and YN configurations was receiving transmitted rays. This caused a small amount of radiation rays to be reflected from the transparent plastic sheet. Therefore, net radiation contribution in this study was based on the transmitted radiation.

The comparison of evaporation rates between each pan is shown in Figure 3. According to Liu and Xia [10], Chu et al [15] and Goyal [16] study, the evaporation flux from the free surface of water increased with the higher air vapor pressure deficit, higher air temperature and higher wind speed.

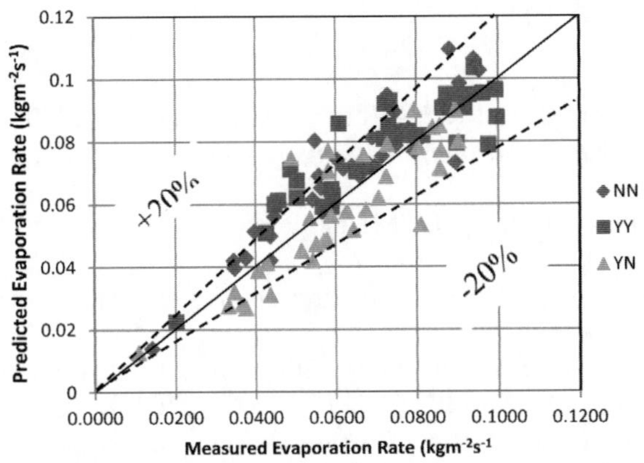

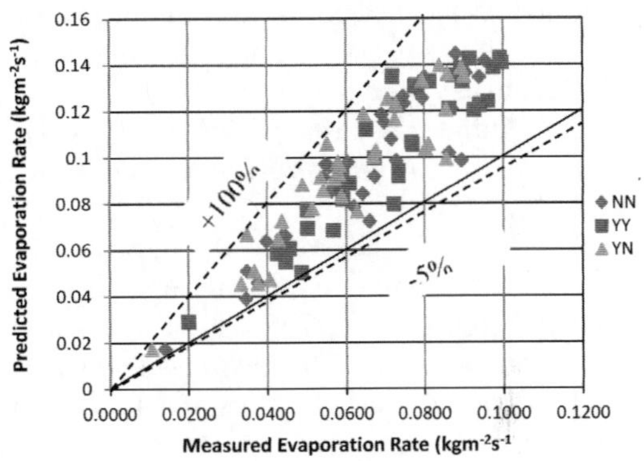

Figure 4: Evaporation Rate estimated by Rayner [5] model (Left) appears to predict all three configurations reasonably while Penman [13] tends to overpredict (Right) contained within the envelope of -5% to +100%.

Figure 4 shows the Penman [13] model apparently over-predicting the evaporation rate for all configurations. Rayner [5] has performed better by far. It seems Rayner model could cope with the aggregated change of characteristics of all three configurations with the minimum percentage difference ranging between 0.6 to 52.4 while Penman was from 2.7 to 91.5. The maximum percentage difference in prediction for both models were for the YN configuration, implying that this was the least stable of all configurations, since flow reversal at the chimney exit was detected by smoke flow visualization and CFD analysis.

The plot of predicted and measured evaporation rates of all three configuration on the same graph in Figure 4 shows that the Rayner [5] model appears to predict the data of all configurations satisfactorily and a recent work Li et al [17] has confirmed its satisfactory performance. While this is helpful for design purposes, the model's accuracy copes with effects and does not explain the cause between the difference of configurations YY and YN if one is not given the knowledge of the different geometrical arrangements.

6. Conclusion

This study assessed the ability of canopied solar irradiated natural ventilation to enhance the evaporation of free surface water in pans data collected under the East Malaysian's hot and humid climate. It also investigated the influence of the meteorological parameters including the net solar radiation on the free surface of water, the vapor pressure deficit, and ability to transfer vapor from the free surface of water. The observation results concluded that:

- The evaporation rate for all three evaporating configurations could be predicted satisfactorily by Rayner [5] but not Penman [13].
- the configuration YY, where natural draft and vapor pressure increased 13% higher evaporation rates than configuration NN, was because as the vapor pressure deficit rises, the water molecules begin moving amount more rapidly.

- the configuration YY, where natural draft and vapor pressure increased 20% higher evaporation rates than configuration YN, was not characterized by any one of thermodynamic or flow parameters but an aggregate of them according to Rayner [5], and possibly due to the increase of surface air velocity and hence reduction of pressure for the same energy input.

Acknowledgements
The authors would like to thank the Ministry of Higher Education for providing a fundamental research grant with number and Universiti Malaysia Sabah for the facility and laboratory support FRG0352-TK-1/2013 to carry out the experimental research.

References

[1] Ahmed, M., Shayya, W.H., Hoey, D., Mahendran, A., Morris, R., and Al-Handaly, J. 2000. *Use of Evaporation Ponds for Brine Disposal in Desalination Plants. Journal of Desalination.* **130**: 155-168.
[2] Jhajharia, D., Agrawal, G., Sevda, M. S. 2009. Influence Of Meteorological Parameters on Pan Evaporation at Agartala. *J. Agric.* Eng. 46(1).
[3] Pereira, H. C. 1957. Field measurements of water use for irrigation control in Kenya coffee. *Journal of Agricultural Science*. **49**: 459-66.
[4] Roderick, M. L, Rotstayn L. D, Farquhar G. D, Hobbins M. T, 2007. On the attribution of changing pan evaporation. *Geophys Res Lett* ;**34**: L17403, doi: 10.1029/2007GL031166.
[5] Rayner DP. 2007. Wind run changes: the dominant factor affecting pan evaporation trends in Australia. *J Climate*. **20**:3379-94.
[6] Liu, X., Luo, Y., Zhang, D., Zhang, M., Liu, C., . 2011. Recent changes in pan-evaporation dynamics in China. *Geophys. Res. Lett.* **38(13)**
[7] Chu C. M., Rahman, M.M. and Kumaresan, S.. 2012. Effect Of Cold Inflow On Chimney Height of Natural Draft Cooling Towers. Nuclear Engineering and Design, *Elsevier*. **249**: 125-131.
[8] Chow, V. T., Maidement, D. R., Mays, L. W. 1988. Applied Hydrology. *McGraw Hill Book* Co., New York.
[9] Shuttleworth, W. J. 1993. Evaporation. IN MAIDMENT, D. R. (Ed.) Handbook of Hydrology. New York, *McGraw-Hill Inc*.
[10] Liu, Q., Xia X. H., 2012. Contribution of meteorological variables to changes in potential evaporation in Haihe River Basin, China. *Procedia Environmental Sciences* **13**: 1836 – 1845
[11] Smith, K., Lavis, M. E. 1975. Environmental influences on the temperature of a small upland stream. *Oikos* **26(2)**: 228-236.
[12] Jackman, A. P., and Yotsukura, N. 1977. Thermal loading of natural streams: U.S. Geol. Survey Prof. Paper 991, 39p.
[13] Penman, H.L., 1948. Natural evaporation from open water, bare soil and grass. *Proc. Royal Soc. London* **A193**:120-145.
[14] Thom, A. S., Thony, J. L, Vauclin, M., 1981. On the proper employment of evaporation pans and atmometers in estimating potential transpiration. Quart. J Roy Met. Soc. **107**: 711-736.
[15] Chu C. R., Li, M. H., Chen, Y. Y., Kuo, Y. H., 2010. A wind tunnel experiment on the evaporation rate of Class A evaporation pan. *Journal of Hydrology*. **381**: 221-224.
[16] Goyal, R. K. 2004. Sensitivity of evapotranspiration to global warming: a case study of arid zone of Rajasthan (India). Agr. Water manage; **69**:1-11.
[17] Li, M., Chu, R., Shen, S., Islam, A. 2018. Dynamic analysis of pan evaporation variations in the Huai River Basin, a climate transition zone in eastern China. Science of the Total Environment **625**, 496–509, *Elsevier*.

IOP Conf. Series: Materials Science and Engineering **606** (2019) 012014 doi:10.1088/1757-899X/606/1/012014

Evaluating the contraction value of ferromagnetic material at early fatigue loading stage using magnetic flux leakage signature

A Arifin*, S N Sahadan and S Abdullah
Centre for Integrated Design for Advanced Mechanical Systems
Faculty of Engineering and Built Environment
Universiti Kebangsaan Malaysia, 43600 UKM Bangi, Selangor

Corresponding author: *azli@ukm.edu.my

Abstract. This work focuses on the investigation to evaluate a material parameter, namely a contraction value ψ, at failure in estimating the fatigue life of ferromagnetic steel based on magnetic metal memory method. It begins with specimen preparation made from mild carbon steel and it was machined as specified in the ASTM E467-01 standard. These cyclic tests were performed using the 25-kN servo-hydraulic machine with the applied axial loading levels were at 60% UTS and 70% UTS. The magnetic flux leakage was then measured at every hour of the load applied, and the nature of change of magnetic leakage intensity H_p over the surface of the specimen. A further analysis on the collected signals or pattern was carried out based on the H_p distribution and its gradient at dH_p/dx, and both parameters were later being used to estimate total loading time before specimen failure. A further analysis on the collected signals or pattern was carried out based on the H_p distribution and its gradient at dH_p/dx, and both parameters were later being used to estimate total loading time before specimen failure. At every hour of loading cycles, the fatigue life of the specimen is estimated using the specific calculation of this MMM approach.

1. Introduction

Critical location of the component made from ferromagnetic materials, such as steel and cast iron can be potentially monitored using magnetic state for their degradation failure process. Magnetic measurements are a Non-Destructive Test (NDT) technique that can be used to evaluate fatigue damage on ferromagnetic materials due to the modification of density and lattice dislocation patterns, and the formation of continuous sliding bands causing changes in magnetic properties. Major magnetic techniques that have the potential to detect fatigue damage include the effect of Barkhausen (Barkhausen Noise), magneto-acoustic emission (Magneto-acoustic Emission), magnetic flux leakage and magnetic hysteresis. Barkhausen Noise and Magneto-acoustic Emission Signals are derived from the non-continuous magnetization process that occurs during the hysteresis cycle, and both are greatly influenced by stresses and deformations [1]. However, the disadvantage of this technique in NDT applications is that the Barkhausen Noise and Magneto-acoustic Emission signal is generally weak and, therefore, highly sensitive measurement equipment is needed and the effect on noise interference is high and the eddy current method, need an external excitation to detect crack.

To overcome disadvantages of previous method, a Russian scholar, introducing Metal Magnetic Memory-MMM. This method is a passive non-destructive testing method developed based on magneto-mechanical effects. It is a method of advanced magnetics flux leakage, in which the mechanism is like the magnetic flux leakage method but does not require artificial excitation. Due to the impact of the load and the earth's magnetic field, spontaneous magnetic signal changes may occur on the surface of the ferromagnetic material during manufacturing or operation [2]. Fracture due to fatigue usually occurs in areas where significant stress concentration zones microscopic exist, which became a key resource in the development of defect. Recent studies have shown that the fatigue cracks occur on ferromagnetic material surfaces are initiated and induced spontaneously fluctuation of magnetic signals.

Using metal magnetic memory methods, the obtained magnetic signal distribution can be used to evaluate the stress concentration conditions that occur and provide an early indication of the potential failure. Based on the concept of metal magnetic memory, two parameters used to measure the free magnetic signals in stress concentration area are normal component signals, $H(y)$ and tangent component signal, $H(x)$ [3]. The stress concentration level on the specimen can be determined by the maximum value of the normal components gradient, $dH(y)/dx$. This $dH(y)/dx$ value has the potential to be an indicator in monitoring fatigue crack growth [4]. Signals $H(x)$ and $dH(y)/dx$ show different characteristics at different stages of growth of cracking. Therefore, the characterization of MMM signals at different levels of fatigue crack growth can be implemented [5]. MMM has shown capability in positioning crack due to cyclic loading [6]. Chen et al. [7] in 2017 has shown defect classification in metal using enhancement MMM method. A development of magneto mechanical model by Shi et al. [8] in 2017 to establish quantitative relationship between the shape and size of defect and the surface of magnetic memory signals. This method also has potential in estimating residual life of a component in service base on the magnetic flux leakage based on static loading condition [9]. (Dubov & Kolokolnikov, 2008).

This paper discussed an experimental research on the Magnetic Metal Memory (MMM) method under uniaxial fatigue loading. The objective of this study is to determine stress concentration zone of the specimen at early stage. After stress concentration being detected, residual life of the specimen is estimated. The materials properties called contraction at failure was evaluated to analyse the effect of contraction value with fatigue life estimated. The estimated fatigue life were than compared with experimental value and analytical calculation. In this study length-based signals were used to monitor signal from uniaxial cyclic testing. There is a lot of spaces can be explored to gain better knowledge relating to this method. Although research relates to metal magnetic memory and fatigue has been carried out, those researches gave more focus on positioning and sizing the crack. By carrying this early stage research towards MMM method life time estimation, it is hopes that this study can channel new method for fatigue life evaluation of a component.

2. Theoretical background

2.1 Fatigue failure

Fatigue is a dangerous phenomenon because it can cause sudden failure due to microscopic scale changes, which is the dislocation of the atomic lattice on the material structure after having been in service for many years [10]. Knowledge of fatigue failures in metal materials, components and structures is less clear and incomprehensible, where it is difficult to predict by engineers due to the complex nature of this failure mechanism [11]. Proper prediction of fatigue is crucial to ensuring component safety and reliability. To control the quality of ferromagnetic materials, it is important to monitor the timidity of fatigue in a timely manner to detect the beginning and propagation of fatigue cracking, and to estimate the remaining life.

The fatigue damage for each loading cycle D_i can be calculated as follows:

$$D_i = \frac{1}{N_f} \tag{1}$$

The Palmgren-Miner rule is then used to calculate the cumulative fatigue damage D of a loading block, and it is defined as follows:

$$D_i = \sum \frac{n_i}{N_f} \tag{2}$$

where n_i is the number of applied cycles. Fatigue damage has a range of zero to one, where zero indicates no damage (infinite cycles to failure) and one is assumed as failure (one cycle to failure) [12].

2.2 Magnetic Metal Memory Method

MMM inspection using natural magnetization and-after effects, which succeeded in the form of magnetic memory of the metal to actual strains and change of metal structure with no preparation operation. As loads applied, ferromagnetic material tends to change the dimension of the materials in a microstructure scale. Magnetic domain change from original condition when load applied. This phenomenon called magneto-mechanical effect, introduced by Jiles in 1995 [13]. It is based on magneto elasticity and magneto-mechanical characteristics of materials. During operation or loaded, the magnetic domain will dislocate and be irreversibly reoriented.

Two parameters used to measure the stress concentration zone are $H_p(x)$ to measure the intensity of the magnetic field parallel to the surface material while $H_p(y)$ to measure the intensity of magnetic field perpendicular to surface material. Figure 1 shows the reaction components of $H_p(x)$ and $Hp(y)$ in the zone stress concentration. $H_p(x)$ will form a peak and the polarity $H_p(y)$ changes from negative to positive at stress concentration zone. The tangent component of the magnetic leakage $H_p(x)$ shows the maximum value, while the normal component value of the magnetic leakage $H_p(y)$ is zero.

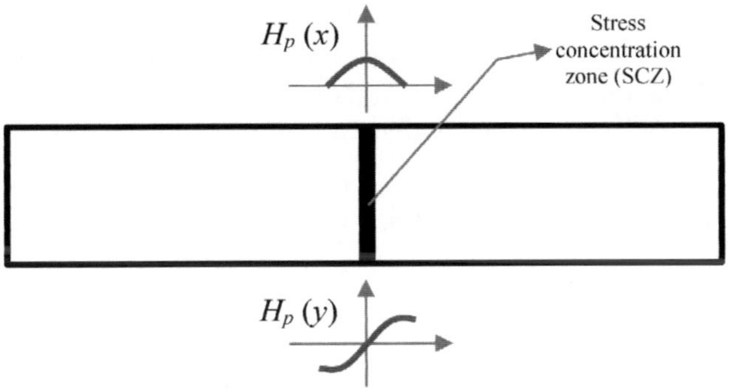

Figure 1: Reaction of component $H_p(x)$ and $H_p(y)$.

2.2.1 Contraction at Failure, ψ.

The mechanical characteristics values depend on dimension and shape of the test specimen as well as the method of measurements. The values of limiting mechanical characteristics for materials embodied in specific shapes of parts or objects may considerably differ from references. Limiting values mechanical characteristics of the same material called extension strain ε_t and contraction ψ at failure. These two characteristics are the characteristics of the plasticity of the materials. Combined with limiting values of conditional specific load or strains, giving characteristics of energy or fracture energy. Correlation between the extension strain and contraction value, making these characteristic interchanges.

Contraction is determined in the neck region and extension strain is an average elongation on a certain base, the dimension of which strongly influences the result-the extension strain amount. To reduce the longitudinal strain assessment error, the concept of the true strain was introduced and determined as $\varepsilon_{true} = \ln(1 + \varepsilon_{long})$. However, for description of plastic deformability of material as a function of strain variation due to conditional load, changing from minimum to destructive values, it is possible and appropriate to use extension strain since this characteristic agreed with other strain characteristics. As for contraction, this characteristic is very important already for failure conditions evaluation and for determination of the role of the very longitudinal and transverse linear strains during failure. Correlation of extension strain and contraction, which with extension strain ε_t and contraction at failure ψ characteristics for various steels and alloys reveal their correlation as

$$\varepsilon_t = f(\psi) \tag{3}$$

2.2.2 Magnetic metal memory life estimation. The parameter in the MMM method is the magnetic leakage field H_p, gradient (dH_p/dx) or field variation intensity factor, K_{in}. Research relation of energy between the magnetic and mechanical parameters of specimen under static and cyclic loads [12] given by Eq. (4),

$$m = \frac{K_{in}^t}{K_{in}^y} = \left(\frac{\sigma_t}{\sigma_y}\right)^2 \tag{4}$$

where K_{in}^t, K_{in}^y, σ_t and σ_y are values of the magnetic field gradient of conditional ultimate strength, gradient of conditional yield strength, ultimate tensile strength and yield strength respectively. The magnetic parameter K_{in} characterised the magnetic energy density W_m conditioned by the mechanical deformation energy due to force action W_f as shown in Eq. (5),

$$K_{in} = \frac{W_m}{W_f} = \frac{2E}{\sigma^2} \tag{5}$$

where E is modulus of elasticity. The energy relation obtained in Equation (2) then proposed to estimate the lifetime in stress concentration zone (SCZ) based on the actual maximum K_{in}^{act} and the actual service hours of a specific unit of the date of inspection, T_{act} and the period of specimen operation, T_{lim} given as in Eq. (6). Therefore, the residual lifetime of the inspected with SCZ estimated by Eq. (7),

$$T_{lim} = \frac{K_{in}^t}{K_{in}^{act}} (T_{act}) \tag{6}$$

$$T_{life} = T_{lim} - T_{act} \tag{7}$$

3. Methodology

In this study, the material used is the ferromagnetic material. Table 1 show the chemical composition of the specimen. The magnetic flux leakage sensor was used in this study. Due to that, steel grade SAE1045 has been chosen because it is ferromagnetic material with ability to induce magnetic flux for crack detection and evaluation. SAE 1045 steel has excellent mechanical properties and widely used as structural component [14].

Table 1. Chemical Composition of SAE 1045 steel (wt. %).

	Composition (wt. %)				
	Fe	C	Mn	P	S
Min	98.51	0.42	0.60	0.010	0.020
Max	98.98	0.50	0.90	0.040	0.050

Tensile test has been carried out to gather the mechanical properties of the specimen. Material has been prepared according to ASTM E8-01 (Standard Test Method for Tension Testing of Metallic Materials). Figure 2 show stress-strain curve generated from tensile test and mechanical properties of the specimen tabulated in Table 2.

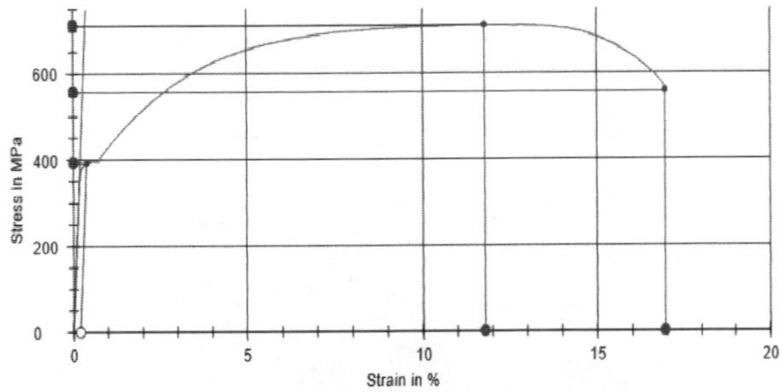

Figure 2: Stress-strain curve of SAE 1045 steel

Table 2. Mechanical properties of SAE 1045.

Material	Mechanical properties		
	Yield Strength (MPa)	Ultimate Strength (MPa)	Young Modulus (GPa)
SAE 1045	392	710	190.5

For the cyclic testing, specimen has been prepared according to ASTM E466-07 (Standard Practice for Conducting Force Controlled Constant Amplitude Axial Fatigue Test of Metallic Materials). Specimen has been polished to mirror like surface to decrease stress concentration that lead to crack initiation during fatigue testing. The gauge parts of the specimen were divided to eight positions with 10 mm distance from point 1 to point 9 as shown in Figure 3. This is to facilitate the detection of stress concentration location after signal processing is performed on the observed magnetic flux leakage signal. Electromagnetic signals observed offline, in which the machine is stopped at an interval of every hour of loading to get the relation between fatigue life of the experiment and fatigue life prediction using the metal magnetic memory and stress concentration changes with increasing time.

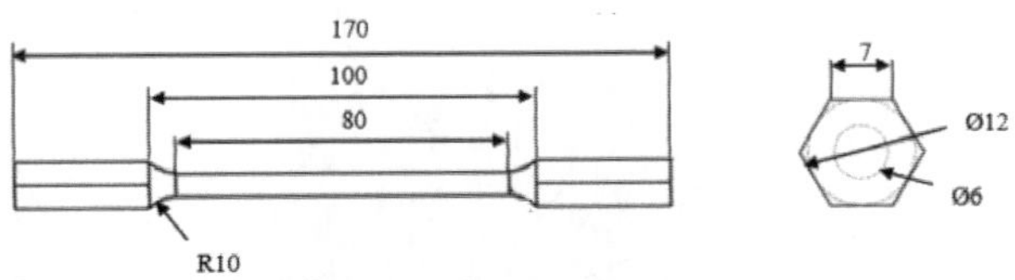

(a)

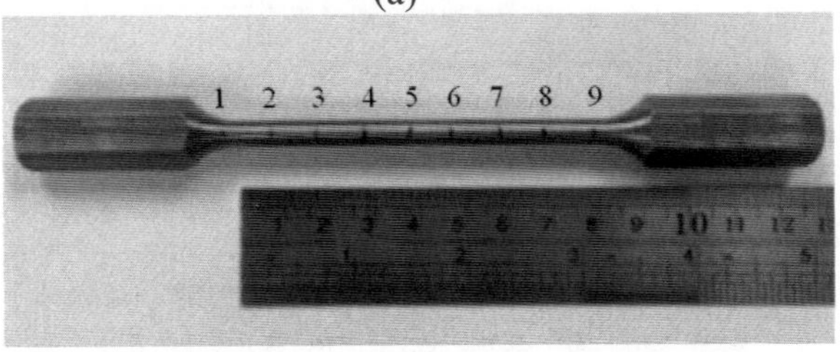

(b)

Figure 3 Specimen for cyclic testing a) specimen draft b) actual specimen used in experiment and point on the gauge length

Purpose of cyclic testing is to determine fatigue life of metallic specimen. The test conducted using a servo-hydraulic machine with capacity of 25 kN in room temperature. Figure 4 shows the servo-hydraulic machine with specimen attached to the machine. The loading capacity used for the uniaxial cyclic testing is 0.7 of ultimate tensile strength (UTS) and 0.75 UTS with frequency of 1 Hz.

IOP Conf. Series: Materials Science and Engineering **606** (2019) 012014 doi:10.1088/1757-899X/606/1/012014

Figure 4: The servo-hydraulic machine used in the study

During the testing, magnetic flux leakage signals are collected using a stress concentration test device. This device is capable of recording data that is a magnetic signal on the memory card. Magnetic changes influenced by stress concentration, can be seen on the display screen on the device in the form of magnetic graphs against the distance as shown in Figure 5. The advantage of this device is its ability to detect stress concentration zones on small structures and equipment that are in a difficult area. The device also features two transducer flux channels and wheel distance calculations. To facilitate the process of observing magnetic signals on specimens, aluminium runways are used. Figure 6 shows how the magnetic signal is observed during the test. Stress concentration test device are connected to a computer data acquisition system to process data in obtaining stress concentration locations due to cracking in the specimen. From the analysis, the residual life of the specimen can be estimated.

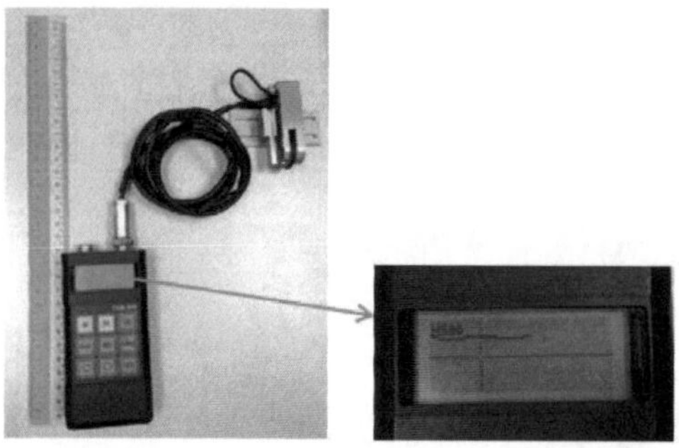

Figure 5: Stress concentration test device and display of the stress concentration graph on screen

Figure 6: Method of observing magnetic signal offline using aluminium platforms

IOP Conf. Series: Materials Science and Engineering **606** (2019) 012014 doi:10.1088/1757-899X/606/1/012014

4. Results and Discussion

Through metal magnetic memory sensing devices used, two magnetic flux leakage signals are obtained namely normal component signals, H(y) and normal component gradient signals, dH (y) / dx. Figure 7 shows the graph of the signal H(y) against the length of the runway track and signal d*H*(*y*)/d*x* against the length of the observation track, *Lx* for 70% UTS cyclic loading. Figure 7 (a) for normal component signals and Figure 7 (b) for gradient of normal component signals. Magnetic signals are shown by combining signals obtained from different cycles of life. Cycle load at 70% UTS produces fatigue life of about 11 hours 36 minutes equivalent to 41760 cycles. Signal was taken at an hour interval. The signal, *H*(*y*) starting from before the clamp, clamping, the first hour until the 11 hours of loading. The lowest value of the *H*(*y*) given by signal before clamping, followed by signal during clamping at servo-hydraulic machine. At the moment of the testing begin, tabulation of stress concentration increases rapidly. This increment due to early stages of fatigue failure, the decrease in *H*(*y*) is due to hardening or softening of the strain, causing changes in the placement structure. In the next fatigue cycle the structure of the dislocation cells becomes stable and the magnetic surface intensity decreases until there is no change. At the last stage micro-crack growth was followed by macroscopic crack propagation resulting in a significant increase in magnetic signals [15].

Analysis by the normal component gradient signals, d*H*(*y*)/d*x* as shown at bottom of Fig. 7(b) is more appropriate in assessing the extent of damage or failure experienced by the specimen [16]. Accordingly, further analysis in this study focuses on the signal change value of d*H*(*y*)/d*x*. In Fig. 7(a), the change shown by signal d*H*(*y*)/d*x* at 0 to 4 mm for all cyclic intervals value is more pronounced than the *H*(*y*) signal. The height of signal curve value of d*H*(*y*)/d*x* shows the gradient value of *H*(*y*) signal. Significant stress concentrations can be seen at 0 to 4 mm with a higher altitude of the signal curve than at other distances. Through the indicated signal there is a high stress concentration at 0 to 4 mm which is potentially a fatigue failure position. Visual examination from the fractured specimen, the fracture position was matched with the high peak given by d*H*(*y*)/d*x* distribution. Position of fracture specimen shows in Figure 8. For 60%UTS cyclic load, gradient signals d*H*(*y*)/d*x* in Figure 9 shows concentration of distribution is at distance of 16 to 22 mm. Many high peak gradient d*H*(*y*)/d*x* can be seen clearly from the figure. This signal show that, at that position is the potential of fatigue failure position.

Value of contraction at failure, ψ for certain loading hours were assessed to study the effect of this value in estimating total loading hours for the specimen. From Table 3 and Table 4, it is found that the value of ψ is vary from 0.33 to 0.8 to achieve similar fatigue life as the experiment. At early loading hours, the maximum value is 0.8 but this value is unable to achieve total service hours as gathered from experiment. As the loading hours increase, the value of ψ is decrease. The optimum contraction value to achieve total loading hours is in between of 0.77 to 0.79 as shown in Table 3 and Table 4 for 60% UTS and 70%UTS cyclic loading respectively. For lower cyclic loading, the contraction, ψ value variation is bigger compared to lower cyclic loading. Estimation of total loading hours shows the influence of contraction value in estimating total loading hours for each specimen. Figure 10 shows the tabulation of the contraction towards specimen failure in time that showing decreasing value of contraction as loading hours increased. The decreasing value in contraction showing the influence of this magnetic signature towards the life of the specimen. Again, it shows that the MMM methods proposed in this study have strong potential in estimating fatigue at early stage loading.

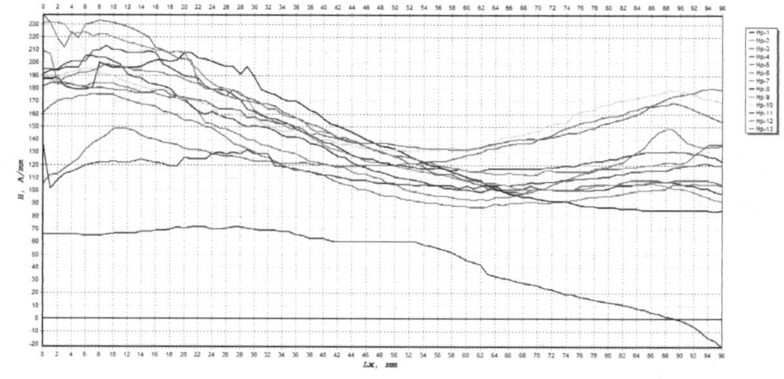

(a)

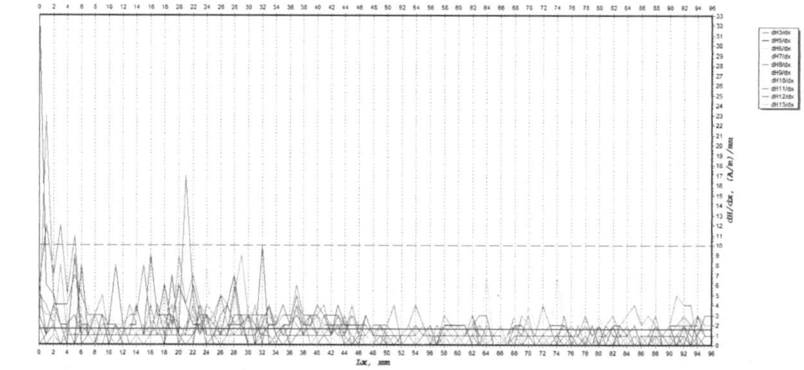

(b)

Figure 7: (a) Distribution of magnetic leakage field intensity, H_p and (b) distribution of magnetic leakage field intensity gradient, (dH_p/dx) along the specimen gauge length for 70% UTS loading.

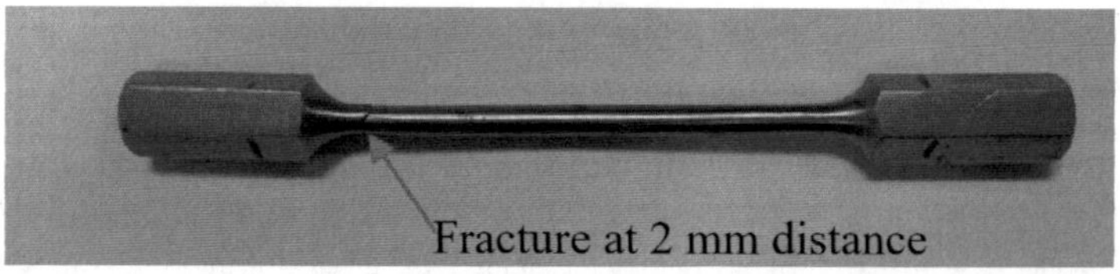

Figure 8: Fracture position on the specimen at 70%UTS cyclic loading

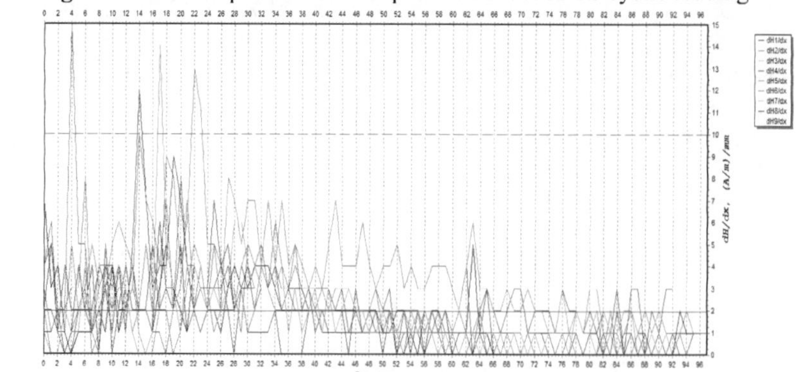

Figure 9: (a) Distribution of magnetic leakage field intensity gradient, (dH_p/dx) along the specimen gauge length for 60% UTS loading.

Table 3: Contraction value for specimen with 60% UTS cyclic loading

Service Hours	Contraction, ψ	Total loading hours estimated using MMM Life
2	0.80	9
3	0.80	13
4	0.80	15
5	0.80	30
6	0.80	32
14	0.79	70
22	0.68	70
30	0.56	70
38	0.56	70
44	0.33	70

Table 4: Contraction value for specimen with 70% UTS cyclic loading

Service Hours	Contraction, ψ	Total loading hours estimated using MMM Life
1	0.80	2
2	0.80	5
3	0.79	11
4	0.75	11
5	0.75	11
6	0.77	11
7	0.67	11
8	0.54	11
9	0.48	12
10	0.41	12
11	0.39	12

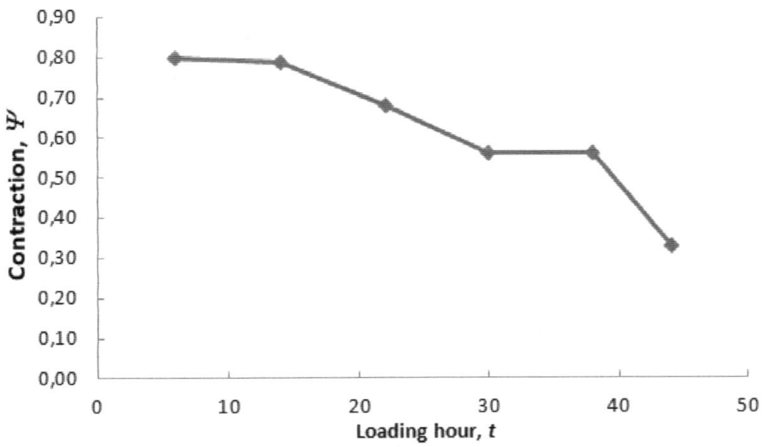

(a)

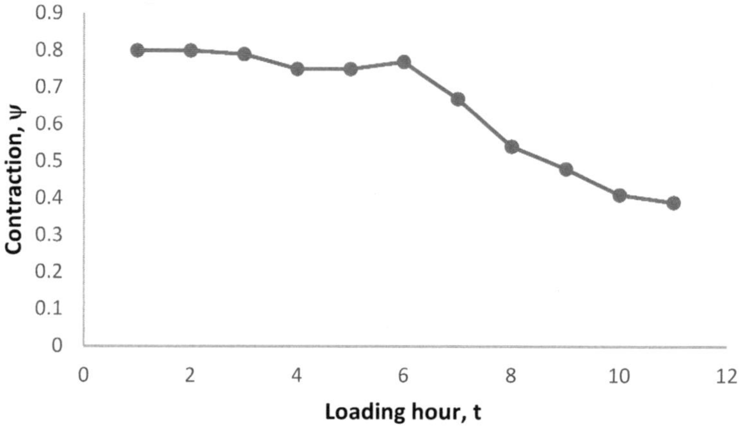

(b)

Figure 10: Tabulation of contraction, ψ value corresponding with loading hour a) 60% UTS cyclic loading b) and 70% UTS cyclic loading

In order to evaluate the ability of contraction value in estimating life of the specimens, the value of life from the MMM estimation were compared with the value of experiment and analytical calculation. Results given by the number of cycles to failure from the experiment and estimated from the magnetic flux signal, that indicate period of specimen operation tabulated in Table 5. From the Table 5 when applied axial load was 70%$_{UTS}$, the specimen failed at 11 hours 36 minutes or 41760 cycles. The estimated from MMM is 17 hours of life. The value of MMM_{Life} have high different compared to experiment, however this value is almost same as given by analytical calculation which is 16 hours and 15 minutes. Both specimens give almost same trend where experimental life give shorter life compared to MMM_{Life} that is estimated using MMM method and analytical calculation. These discrepancies due actual condition of the specimen that degradation due to manufacturing and processing. However, this estimation shows the potential of MMM method in estimating the fatigue life at early condition before crack.

Table 5: Fatigue life estimation based on MMM method with comparison to experiment and analytical calculation

Loading	Fatigue Life		
	Experiment	MMM Life Prediction	Analytical
60%$_{UTS}$	62 hours and 23 minutes 224580 cycles	70 hours	97 hours and 30 minutes 351000 cycles
70%$_{UTS}$	11 hours and 36 minutes 41760 cycles	12 hours	16 hours and 15 minutes 58500 cycles

5. Conclusion

The evaluation of magnetic flux leakage signature through contraction value at early fatigue loading have shown the capability to estimate fatigue life as similar as value given by the experiment. The gradient of the stress distribution signals $dH(y)/dx$ shows the potential fracture position in the specimen. The density of the stress concentration indicate there is defect on that specific position. Evaluation with fractured specimen has confirmed the position with high stress concentration zone (SCZ) tabulated by signal $dH(y)/dx$ was the fractured position. The materials value evaluated in this study, the contraction value shows it relation in estimating the total loading hours before fracture of the specimen. The value varies from 0.33 to 0.8, and optimum value to estimate life is in between 0.77 to 0.79. The estimated value of life using signals gathered from the MMM sensors, MMM_{life} has small differences compared to value of life from the experiment. At cycles to failure of 41760 cycles which equal to 11 hours 36 minutes, the life estimated from the MMM sensors is 17 hours and at cycles to failure of 224580 cycles which equal to 62 hours 23 minutes, the life estimated using MMM method is 70 hours. The estimated value using the MMM method proposed in this study fall within actual number of cycles by experiment and analytical method. This finding shows the capability this method to locate the position of fatigue crack at early stage, as well as estimating fatigue life of the components.

Acknowledgments
Authors wishing to acknowledge Universiti Kebangsaan Malaysia for financing this study.

Reference

[1] Lo C C, Tang F, Shi Y, Jiles D C and Biner S B 1999 Monitoring fatigue damage in materials using magnetic measurement techniques *Journal of Applied Physics* **85** 4595-97

[2] Tomáš I, Kovářík O, Vértesy G and Kadlecová J 2014 Nondestructive indication of fatigue damage and residual lifetime in ferromagnetic construction materials *Measurement Science and Technology* **25(6)** 065601

[3] Kang C, Cui L, Zhang J, Gao L and Xu Y 2011 Experiment research on the metal magnetic memory in gear micro crack detection. *IEEE International Conference on Mechatronics and Automation* 1163–1167

[4] Huang H, Jiang S, Liu R and Liu Z 2014 Investigation of magnetic memory signals induced by dynamic bending load in fatigue crack propagation process of structural steel *Journal of Nondestructive Evaluation* **33** 407–412

[5] Chongchong L, Lihong D, Haidou W, Guolu L and Binshi X 2016 Metal magnetic memory technique used to predict the fatigue crack propagation behavior of 0.45%C steel *Journal of Magnetism and Magnetic Materials* **405** 150–157

[6] Azli Ariffin, Meor Iqram Meor Ahmad, Shahrum Abdullah and Wan Zulhelmi Wan Jusoh 2015 Detection of crack position due to cyclic loading for ferromagnetic materials based on magnetic memory method *Jurnal Teknologi* **75-7** 67-70

[7] Chen H, Wang C and Zuo X 2017 Reseach on methods of defect classification based on metal magnetic memory *NDT&E International* **92** 82-87

[8] Shi P, Jin K, and Zheng X 2017 A magnetomechanical model for magnetic memory method *International Journal of Mechanical Science* **124-125** 229-241

[9] Dubov A and Kolokolnikov S 2008 Technical Diagnostics of Equipment and Constructions with Residual Life Assessment Using the Method of Metal Magnetic Memory *17th World Conference on Nondestructive Testing*

[10] Božić Ž, Schmauder S, Mlikota M and Hummel M 2014 Multiscale fatigue crack growth modelling for welded stiffened panels *Fatigue & Fracture of Engineering Materials & Structures* Vol. **37(9)** 1043-1054

[11] Ye X W, Su Y H and Han J P 2014 A State-of-the-Art Review on Fatigue Life Assessment of Steel Bridges *Mathematical Problems in Engineering*

IOP Conf. Series: Materials Science and Engineering **606** (2019) 012014 doi:10.1088/1757-899X/606/1/012014

[12] Upadhyaya Y S and Sridhara B K 2012 Fatigue Life Prediction: a Continuum Damage Mechanics and Fracture Mechanics Approach *Materials and Design* Vol. **35** 220-224
[13] Jiles D C 1995 Theory of magneto-mechanical effect *Journal Physic and Applied Pyhsics* Vol. **28** 1537-1546
[14] Xing H, Xu M, Wang R and Zhang J 2006 MMM Fatigue damage evaluation and life prediction modelling for ferromagnetic materials *Key Engineering Materials* **324-325** 619-622
[15] Zambrano O A, Coronado J J and Rodríguez S A 2014 Failure Analysis of a Bridge Crane Shaft *Case Studies in Engineering Failure Analysis* Vol **2** 25-32.
[16] Wang H and Wang J 2016 Simulation of biaxial wheel test and fatigue life estimation considering the influence of tire and wheel camber *Advances in Engineering Software* Vol. **92** 57-64
[17] Leng J, Xu M, Zhao S and Zhang J 2009 Fatigue damage evaluation on ferromagnetic-materials using magnetic memory method *Proceedings of 2009 International Conference on Measuring Technology and Mechatronics Automation* 784-786

IOP Conf. Series: Materials Science and Engineering **606** (2019) 012015 doi:10.1088/1757-899X/606/1/012015

Response Surface Methodology (RSM) based analysis for tool life optimization in cryogenic CO_2 milling of Inconel 718

N H Abdul Halim[2, 3], C H Che Haron[1, 2], J Abdul Ghani[1, 2], M.F. Azhar[2]

[1]Programme of Mechanical Engineering,
[2]Centre for Engineering Materials and Smart Manufacturing (MERCU)
Faculty of Engineering and Built Environment, Universiti Kebangsaan Malaysia, 43600 Bangi, MALAYSIA
[3]Faculty of Mechanical Engineering, Universiti Teknologi MARA, 40450 Shah Alam, MALAYSIA

Abstract. This study presents the tool life optimization of carbide coated ball nose milling inserts using high-speed milling Inconel 718 under cryogenic CO_2 environment. A total of 29 experiments were conducted based on Box-Behnken Response Surface Methodology using the 4 factors; cutting speed (Vc: 120-140 m/min), feed rate (Fz: 0.15-0.25 mm/tooth), axial depth of cut (ap: 0.3-0.7 mm), and radial depth of cut (ae: 0.2-0.6 mm) at three levels each. By using ANOVA, it was found that all factors have significant effects on tool life. However, the feed rate shows its highest influence on the tool performance where the tool life can be lengthened by reducing the feed rate values. After that, the relevant influencing parameters on tool life were derived in the form of mathematical model which was then used to predict the response within the predetermined parameters. With an average error of 8.2% between the predicted and actual tool life, it confirms the adequacy of the model in this research. ANOVA also suggests a tool life of 23.40 min could be obtained at optimized parameters of; Vc: 120.06 m/min, Fz: 0.15 mm/tooth, ap: 0.66 mm and ae: 0.53 mm under cryogenic CO_2 condition.

1. Introduction

Inconel 718 is a nickel-based alloy and it is the most widely used super-alloy for fabricating components operating in an aggressive service environment in heavy industries such as aerospace, marine and petrochemical plant. Inconel 718 is selected to apply because it has advanced mechanical and chemical properties that allow it to withstand its properties at stringent operating conditions. In aerospace industry, it is specifically used to produce the hottest components of turbine engine such as turbine blades which operates up to 520°C of service temperature. It is well performed at such elevated temperature due to its high heat and corrosion resistance and high hot hardness and strength [1]. Moreover, this difficult-to-cut material also has good tensile, creep and rupture strength that upsurges its demands in industries [2]. However, those superior properties develop more challenges in terms of its machinability, particularly when machined in high speed. For Inconel 718, its high-speed machining begins at speed approximately over 50 m/min.

Rapid tool wears that shorten the tool life is among the most difficulties that need to be tackled when machining Inconel 718 in high speed. As reported by researchers such as Patil et al.[3], Aramcharoen et al. [4], Nalbant et al. [5] and Hadi et al. [6], the major cause of short tool life is due to the advanced mechanical and chemical properties of Inconel 718 that increase mechanical and thermal loads on the cutting tool. As consequences, the tool lost its hardness, leading to severe plastic deformation and wear acceleration. Therefore, proper selections of cutting parameters such as cutting speed, feed rate and depth of cut as well as cutting coolant are important as they have significant influences on tool performances. For instance, Kasim et al.[7]asserted that cutting speed, feed rate, and depth of cut provided significant effect on tool life during high-speed milling Inconel 718 in Minimum Quantity Lubrication (MQL) condition. They also found that longer tool life can be achieved at lower values of those parameters. This finding was supported by Abdul Hadi [8] who also milled Inconel 718 at high speed but in cryogenic machining using liquid nitrogen (LN_2). According to their report, higher cutting speed increased the cutting temperature and resulted in severe tool wear. It is because, at high cutting temperature, Inconel 718 tends to harden rapidly that increases its hardness as well as its difficulty to be machined. It happens when its poor thermal conductivity that causes heat which is generated during the shearing of material fails to dissipate through the workpiece and the chips but accumulates at the cutting edges [9]. The heat builds up to the extreme values and then softens the cutting tool materials and accelerates wear. Besides that, Badroush et al. [10] conducted high speed turning Inconel 718 in cryogenic LN_2 condition and found cutting speed and feed rate as the most significant factors affecting the tool life.

To reduce the cutting temperature, the application of cutting fluids in the machining processes is very important. It is used to remove the heat as early as possible and at the same time, to reduce the cutting friction between the cutting tool and the workpiece. The conventional cutting fluid has been proven to be less effective in terms of decreasing the cutting temperature and increasing tool life. Apparently, the application of gasses as a cooling agent in cryogenic machining is seen to be the best option available to counter those limitations. The one that has attracted researchers' attention nowadays is the application of cryogenic machining using carbon dioxide (CO_2). The effects of CO_2 as a coolant agent in machining have shown a lot of benefits to the users as revealed by Machai et al. [11]; decrease notch and flank wear of the tools while machining of titanium alloys, Supekar et al. [12]; provide better lubrication, and Dilip et al. [13]; reduced cutting temperature around 5% to 22% compared to wet machining of turning AISI 1045 steel. Another benefit of CO_2 as reported by Dilip et al. [14] is that it offered a better resistance to rake and flank wear due to the efficient temperature control and the increase of the tool's hardness. Cryogenic CO_2 also has a greater impact on the environment and health and safety issues with major reduction of environmental pollution and workplace hazard.

In this study, the high-speed milling experiments were conducted to cut Inconel 718 in cryogenic condition according to machine parameters designed by Box-Behnken Response Surface Methodology (RSM) method. Then, statistical analysis using ANOVA was conducted to identify the relationship between controlled factors (cutting speed, feed rate, axial depth of cut (DOC) and radial DOC) on tool life. At the same time, a predictive model for the tool life was developed to predict the variation of tool life within the predetermined parameters. The RSM is a collection of statistical and mathematical techniques useful for developing, improving, and optimizing processes [15]. This technique is more accurate, economical and faster than the traditional method such as 'trial and error', 'listen to noise' or 'based on vibration' practices [16]. Researchers such as Kasim et al. [7], Badroush et al. [10] and Aruna et al. [17] are among RSM users who have applied this technique for optimizing and predicting machining performance.

2. Methodology

In this study, the workpiece used for conducting the experiment is Inconel 718, grade AMS 5663 at measured hardness of 42 ± 2 HRC on its top surface. It underwent double aged heat treatment from its raw grade of AMS 5662 with the hardness of 92 HRB according to the material supplier. The dimension of the workpiece is 170 mm x 100 mm x 50 mm (length x width x height) in the form of a solid block. The chemical composition by weight percentage of this commercially available Inconel 718 contains; Ni (53%), Cr (18.3%), Fe (18.7 %), Nb (5.05%), Mo (3.05%), Ti (1.05%), Al (0.49%),

IOP Conf. Series: Materials Science and Engineering **606** (2019) 012015 doi:10.1088/1757-899X/606/1/012015

Co (0.3%), Mn (0.23%), Si (0.08%), C (0.051%), Cu (0.04%), B (0.004%), P (<0.005%) and S (<0.002%).

For the cutting tool, Sumitomo carbide coated ball nose milling inserts with grade ACK 300 were used. It is a multi-coated tool by the alternate layers of TiAlN and AlCrN until it achieves 3 µm of coating thickness. The radial rake angle, axial rake angle, and approach angle are 0°, -3°, and 90° respectively. The insert is 10 mm diameter attached to a 16 mm diameter tool holder at an overhang of 60 mm. The down-end milling experiments were performed on a CNC milling machine model DMG 635, V Eco having a maximum spindle speed of 80,000 rpm. The independent factors for the experiment were cutting speed (Vc), feed rate (Fz), axial DOC (ap) and radial DOC (ae) controlled at three levels as listed in Table 1. A total of 29 experiments were conducted in cryogenic CO_2 condition as in Figure 1. The experimental layout was designed using Box-Behnken RSM where all parameters were arranged without combining their highest and lowest value of each factor at the same time. This is to avoid cutting the material under extreme conditions which would probably produce inadequate results [18].

For the cryogenic system, the coolant is the mixture of liquid and gas CO_2 and compressed air which were mixed together in an in-house designed of CO_2 regulator system. Each of them was controlled at pressure 11 bar, 6 bar, and 4 bar for liquid CO_2, gaseous CO_2, and compressed air, respectively. The minimum temperature of the cryogen existed from the nozzle was controlled at approximately -55 °C. The coolant was supplied through a copper nozzle directed to the tool-chip interface in the cutting zone. The CO_2 regulation system was employed to control the pressure that can influence the liquid-gas-snow phase changes which occur along the flow of the coolants in the insulated hose and the nozzle.

The experiment focused on measuring the progress of tool wear and the total cutting time required to complete all pass in order to identify the tool life. Due to that, the cutting process was interrupted after every specific pass interval to observe and measure the growth of wear on the flank face of the insert. The average (VB) and maximum flank wear (VBmax) were measured and recorded by using a three-axis Mitutoyo toolmaker's microscope at 30x of magnification as shown in Figure 2. This microscope was attached with a digital micrometre that can measure in x-axis and y-axis in 0.001 mm resolution. A custom holding jig was used to tightly hold the insert along the measurement process so that the measurement can be accurately performed. The cutting process then was continued using the same insert until the VBmax reached 0.2 mm. Once it reached the limit, the cutting was stopped and the insert was isolated.

Table 1. Cutting parameters and their levels of control

Control parameters	Level of control parameters		
	-1	0	1
Cutting speed, Vc (m/min)	170	190	210
Feed rate, Fz (mm/tooth)	0.15	0.2	0.25
Axial depth of cut, ap (mm)	0.3	0.5	0.7
Radial depth of cut, ae (mm)	0.2	0.4	0.6

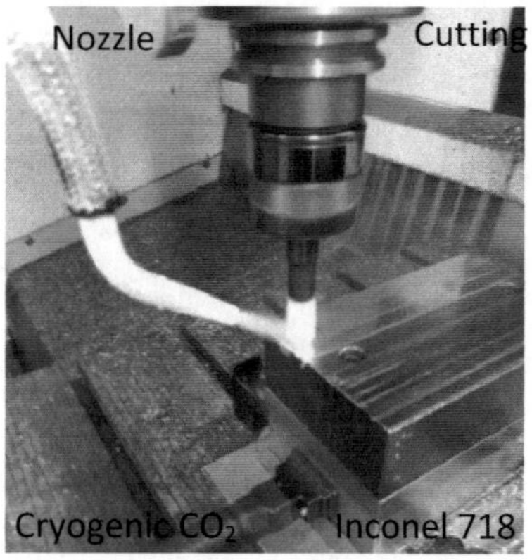

Figure1. Experimental setup with cryogenic CO_2 delivery, and the insert attached to the clamping chuck

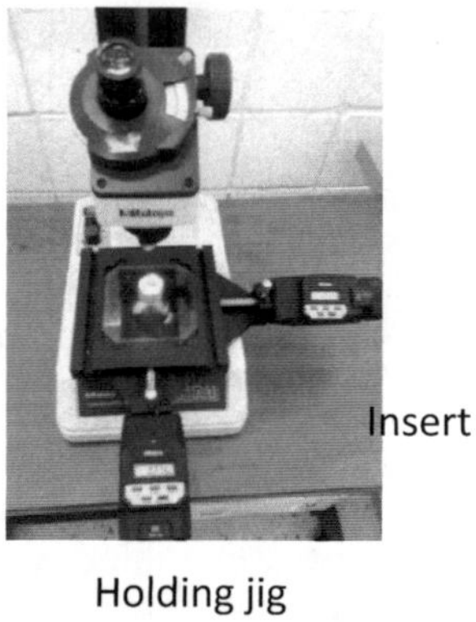

Figure 2. The three-axis Mitutoyo toolmaker's microscope with the insert held by a holding jig

3. Results and optimization

3.1. Tool life from experimental results

Table 2 shows 29 experimental layouts based on Box-Behnken RSM with the experimental and predicted tool life results. Based on the experimental results, the longest tool life was 22.77 min which was recorded at Vc: 120 m/min, Fz: 0.2 mm/tooth, ap: 0.5 mm and ae: 0.2 mm. Meanwhile, the shortest tool life was 3.17 min, obtained at Vc: 130 m/min, Fz: 0.25 mm/tooth, ap: 0.5 and ae: 0.6 mm.

Table 2. Tool life results from actual and predicted

No	Vc (m/min)	Fz (mm/tooth)	ap (mm)	ae (mm)	Tool life (min)		Error %
					Actual	Predicted	
1	120	0.15	0.50	0.40	20.91	20.57	1.62
2	140	0.15	0.50	0.40	10.26	11.31	10.24
3	120	0.25	0.50	0.40	11.39	11.11	2.48
4	140	0.25	0.50	0.40	6.00	7.11	18.43
5	130	0.20	0.30	0.20	17.63	19.30	9.47
6	130	0.20	0.70	0.20	13.47	13.36	0.82
7	130	0.20	0.30	0.60	12.8	12.63	1.36
8	130	0.20	0.70	0.60	7.05	6.69	5.16
9	120	0.20	0.50	0.20	22.77	22.44	1.46
10	140	0.20	0.50	0.20	16.9	15.81	6.48
11	120	0.20	0.50	0.60	14.15	15.76	11.40
12	140	0.20	0.50	0.60	7.85	9.13	16.33
13	130	0.15	0.30	0.40	17.58	16.12	8.31
14	130	0.25	0.30	0.40	9.41	9.28	1.33
15	130	0.15	0.70	0.40	9.43	10.18	7.95
16	130	0.25	0.70	0.40	3.77	3.34	11.28
17	120	0.20	0.30	0.40	15.09	14.74	2.29
18	140	0.20	0.30	0.40	17.43	16.24	6.81

19	120	0.20	0.70	0.40	17.25	16.93	1.83
20	140	0.20	0.70	0.40	3.33	2.17	34.74
21	130	0.15	0.50	0.20	15.43	15.43	0.01
22	130	0.25	0.50	0.20	17.36	17.23	0.77
23	130	0.15	0.50	0.60	18.50	17.39	6.01
24	130	0.25	0.50	0.60	3.17	1.92	39.32
25	130	0.20	0.50	0.40	8.91	9.73	9.23
26	130	0.20	0.50	0.40	9.13	9.73	6.59
27	130	0.20	0.50	0.40	9.60	9.73	1.38
28	130	0.20	0.50	0.40	8.46	9.73	15.04
29	130	0.20	0.50	0.40	9.83	9.73	1.00

3.2. Statistical analysis using ANOVA

Statistical analysis using ANOVA was conducted on the experimental results to identify the significant influence on the tool life by the control factors. A second-order statistical equation model was also developed by using the experimental tool life results and is shown in Equation (1). Based on the ANOVA results as in Table 3, the F-Value of the model is 72.72 and the Prob>F is obviously smaller than 0.05. These imply that the second-order polynomial model generated is significant. By arranging the F values in a descending order, the Box-Behnken quadratic model reveals that factors B, D, A, C, BD, D^2, AC, A^2, and AB are the significant model terms in the model. All of them have a significant effect on the performance of the multi-coated tool life with the major influence coming from feed rate. In this research, the tool life can be lengthened by reducing the feed rate values. Table 3 also shows that the lack of fit is 0.0717 which indicates it is not significant. Thus, the model is adequate and fit for predicting the variation of tool life within the predetermined cutting parameters. It can also be noticed that the determination coefficient (R^2) of the model is 0.9668, which specifies that the model adequately represents the real relationship between the control factors. An adequate precision of 32.183 which is greater than 4 indicates an adequate signal, which is considered desirable.

The interaction effect between control factors on the tool life was analysed using 3D surface model graph as shown in Figures 3, 4 and 5. Figure 3 shows the interaction between feed rate and radial DOC. Both approaches which are by increasing the radial DOC at minimum feed rate (0.15 mm/tooth) or by increasing the feed rate at minimum radial DOC (0.2 mm) could be applied to extend the tool life. Based on the interaction between cutting speed and axial DOC as shown in Figure 4, the longest tool life could be achieved either by increasing the cutting speed at the lowest axial DOC (0.3 mm) or increasing the axial DOC when cutting speed is at the minimum value (170 m/min). For interaction between cutting speed and feed rate as in Figure 5, it recommends that the lowest cutting speed (170 m/min) and feed rate (0.15 mm/tooth) should be applied to lengthen the tool life to the longest. However, increasing the feed rate at any value of cutting speed will reduce the tool life. The steepest skewness of feed rate shows its higher influence on the tool performance. This result was similar with the finding of Kasim [7] which investigated high-speed milling Inconel 718 in MQL condition.

Table 3. ANOVA table for tool life

Source	Sum of squares	DF	Mean square	F Value	Prob > F	
Model	771.07	9	85.67	72.72	< 0.0001	Significant
A-Cutting speed	131.94	1	131.94	111.98	< 0.0001	
B-Feed rate	140.15	1	140.15	118.95	< 0.0001	
C-Axial DOC	105.85	1	105.85	89.84	< 0.0001	
D-Radial DOC	133.60	1	133.60	113.39	< 0.0001	
A^2	53.80	1	53.80	45.66	< 0.0001	

D^2	73.38	1	73.38	62.28	< 0.0001	
AB	6.92	1	6.92	5.87	0.0256	
AC	66.10	1	66.10	56.10	< 0.0001	
BD	74.48	1	74.48	63.21	< 0.0001	
Residual	22.39	19	1.18			
Lack of Fit	21.19	15	1.41	4.74	0.0717	not significant
R-Squared	0.9718					
Adj. R-Squared	0.9584					
Pred. R-Squared	0.9228					
Adeq. Precision	32.183					

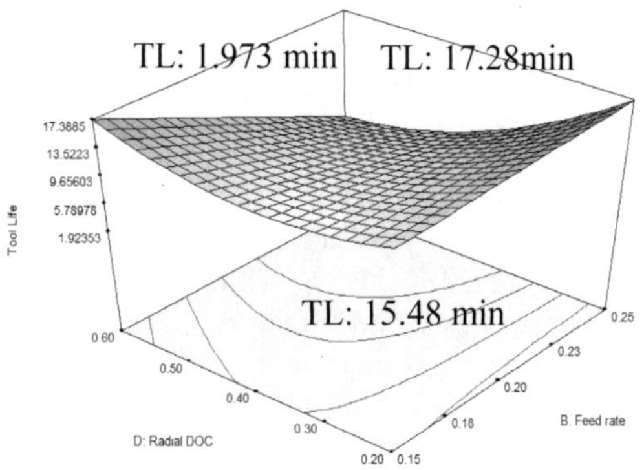

Figure 3. Interaction between Fz and ae at Vc: 130 m/min and ap: 0.50 mm

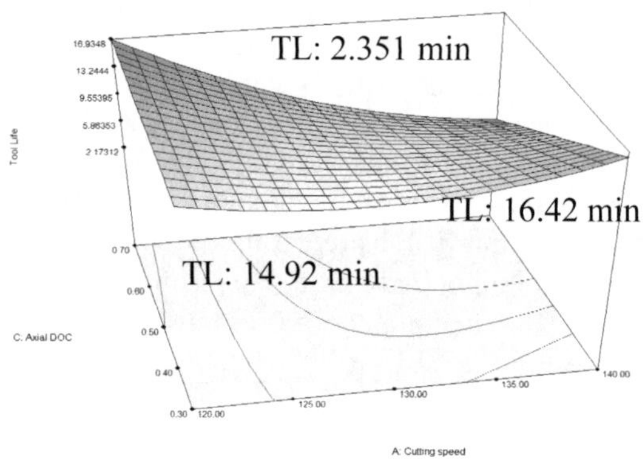

Figure 4. Interaction between Vc and ap at Fz: 0.2 mm/tooth and ae: 0.40 mm

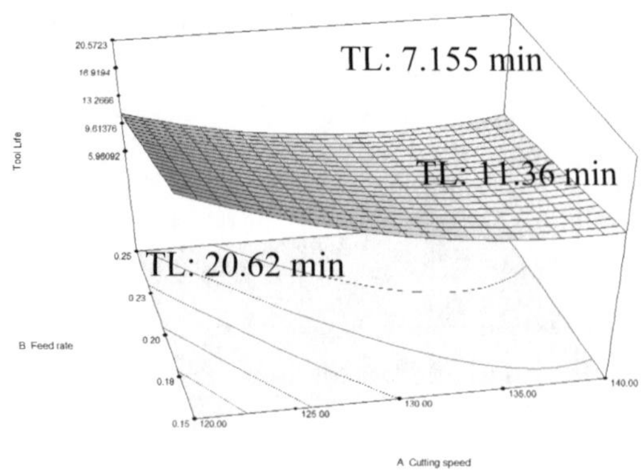

Figure 5. Interaction between Vc and Fz at ap: 0.5 mm and ae: 0.40 mm

3.3. Development and validation of the Second-Order Model using RSM

Second order model equation for the tool life of carbide coated ball nose milling inserts was established using Box-Behnken RSM design as:

Tool Life, TL =

$$252.91894 - 2.57297Vc - 145.60000Fz + 178.23750ap + 4.40360ae + 6.97976E\text{-}003Vc^2 + 81.51634ae^2 + 1.31500Vc.Fz - 1.01625Vc.ap - 431.50000Fz.ae \quad (1)$$

The model was then used to calculate the predicted tool life. The results were compared with the actual experimental results to determine the accuracy of the model. Based on comparison results shown in Table 2 and Figure 6, the errors ranged from 0.01% to 39.32% and the average error was 8.2%. Based on the common practice, the optimization model can be considered valid and acceptable when the error is less than 15%. Moreover, most of the measured tool life was obtained experimentally and the predicted value from equation (1) generated close results.

To validate the adequacy of the mathematical model, the diagnostic plots of the predicted vs. actual and the outlier T as per Figure 7a and 7b were analysed. Both plots proved that the residuals followed a normal distribution and fit the data well. As per Figure 7 (a), all actual and predicted points fall close to the straight line, which means that the errors are distributed normally. Meanwhile, in Figure 7 (b), all data are within the limits, which show no obvious pattern and unusual structure in the model.

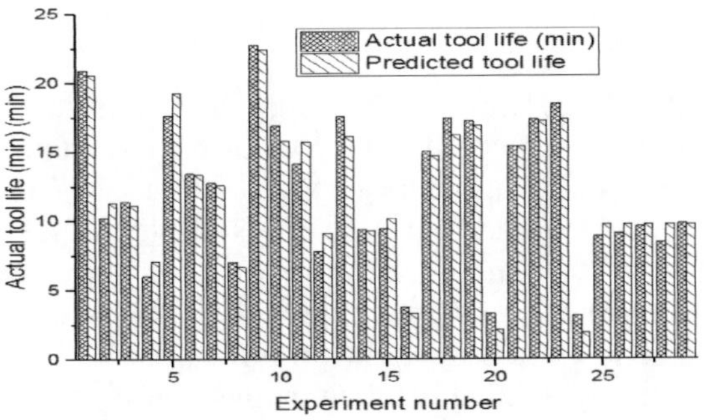

Figure 6. Comparison between actual and predicted tool life of multi-coated carbide inserts

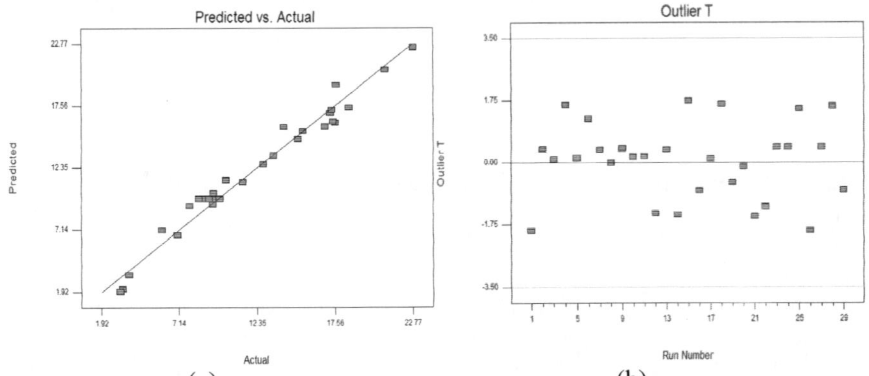

Figure 7. a) The diagnostic plots of the predicted vs. actual, b) Outlier T

3.4. Optimization of tool life

The main objective of the optimization is to determine a combination of optimum processing parameters within the predetermined parameters that can lengthen the tool life to the maximum. ANOVA suggests it can be reached at Vc: 120.06 m/min, Fz: 0.15 mm/tooth, ap: 0.66 mm and ae: 0.53 mm to obtain a tool life of 23.4 min. The ramp view in Figure 8 clearly indicates the point of optimum parameter and the longest tool life in the set parameters. This predicted tool life is longer than the longest tool life that was achieved through the experimental works.

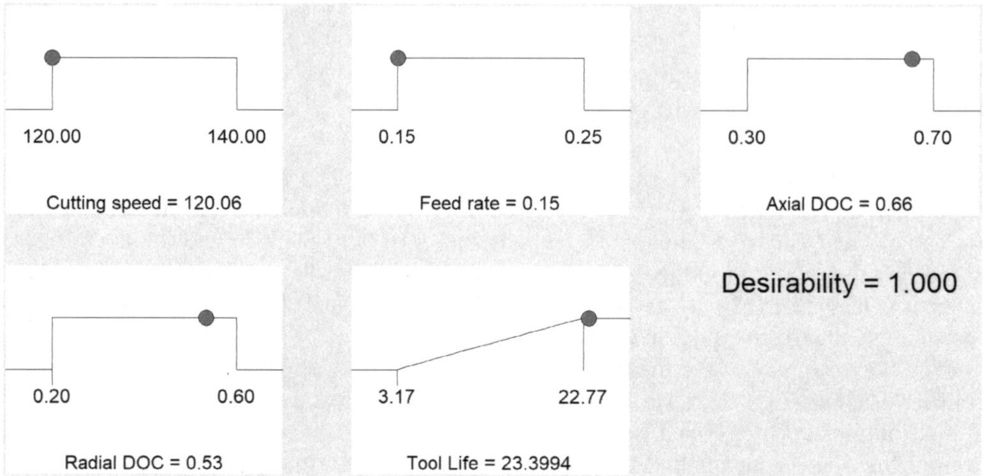

Figure 8. Ramps view of optimum parameters and predicted tool life

4. Conclusion

In this study, the Box Behnken RSM was systematically applied to improve the tool life of carbide coated ball nose milling inserts in high-speed milling of Inconel 718 under cryogenic CO_2 condition. Results from the experimental found that the longest tool life of 22.77 min was achieved at Vc: 120 m/min, Fz: 0.2 mm/tooth, ap: 0.5 mm and ae: 0.2 mm. However, ANOVA suggests the tool life of 23.4 min can be reached at Vc: 120.06 m/min, Fz: 0.15 mm/tooth, ap: 0.66 mm and ae: 0.53 mm. All four controlled factors have influenced on the tool life with the feed rate and radial DOC as the major contributors. For interaction between factors, the main influencer is the interaction between feed rate and radial DOC followed by the interaction between cutting speed and axial DOC. It was also found that the developed mathematical model accurately represented the tool life at an average error of 8.2% when compared to the actual and predicted tool life. This shows that the model is adequate for predicting the tool life values within the range of predetermined parameters in the cryogenic CO_2 environment.

Acknowledgments
Acknowledgement is given to the Government of Malaysia and Universiti Kebangsaan Malaysia for providing the research fund (Grant No: UKM-DIP-2017-023), tools, material, equipment, and support with the technical solutions.

References

[1] D. Ulutan and T. Ozel 2011 Machining induced surface integrity in titanium and nickel alloys : A review Int. *J. Mach. Tools Manuf.* vol. 51 no. 3 pp. 250–280.
[2] E. . Ezugwu, J. Bonney, and Y. Yamane 2003 An overview of the machinability of aeroengine alloys J. Mater. *Process. Technol.* vol. 134 no. 2 pp. 233–253.
[3] N. G. Patil, A. Asem, R. S. Pawade, D. G. Thakur, and P. K. Brahmankar 2014 Comparative Study of High Speed Machining of Inconel 718 in Dry Condition and by Using Compressed Cold Carbon Dioxide Gas as Coolant *Procedia CIRP* vol. 24 pp. 86–91.
[4] A. Aramcharoen and S. K. Chuan, An Experimental Investigation on Cryogenic Milling of Inconel 718 and its Sustainability Assessment *Procedia CIRP* vol. 14 pp. 529–534.
[5] M. Nalbant, A. Altın, and H. Gökkaya 2007 The effect of cutting speed and cutting tool geometry on machinability properties of nickel-base Inconel 718 super alloys *Material and Designs* vol. 28 no. 4 pp 1334-1338.

[6] M. A. Hadi, J. A. Ghani, C. H. Che Haron, and M. S. Kasim 2014 Investigation on wear behavior and chip formation during up-milling and down-milling operations for inconel 718 J. Teknol. *(Sciences Eng.)* vol. 66 no. 3 pp. 15–21.

[7] M. S. Kasim, C. H. Che Haron, J. A. Ghani, A. I. Gusri, M. Z. A. Yazid, and M. A. Sulaiman 2013 Tool life of TiAlN PVD coated carbide tool in high-speed end milling of untreated Inconel 718 under minimum quantity lubrication condition Sains Malaysiana vol. 42 no. 12 pp. 1721–1726.

[8] M. Abdul Hadi, J. A. Ghani, C. H. Che Haron, and M. S. Kasim 2016 Effect of cutting speed on the carbide cutting tool in milling Inconel 718 alloy J. Mater. Res. vol. 31 no. 13 pp. 1885–1892.

[9] K. Busch, C. Hochmuth, B. Pause, A. Stoll, and R. Wertheim 2016 Investigation of Cooling and Lubrication Strategies for Machining High-temperature Alloys *Procedia CIRP* vol. 41 pp. 835–840.

[10] N. Badroush, C. H. C. Haron, J. A. Ghani, M. F. Azhar, N. Hayati, and A. Halim 2018 Performance of Coated Carbide Tools when Turning Inconel Alloy 718 under Cryogenic Condition using RSM J. of Mechanical Engineering vol. 5 no. 3 pp 73–87.

[11] C. Machai and D. Biermann 2011 Machining of β-titanium-alloy Ti–10V–2Fe–3Al under cryogenic conditions: Cooling with carbon dioxide snow *J. Mater. Process. Technol.* vol. 211 no. 6 pp. 1175–1183.

[12] S. D. Supekar, A. F. Clarens, D. A. Stephenson, and S. J. Skerlos 2012 Performance of supercritical carbon dioxide sprays as coolants and lubricants in representative metalworking operations *J. Mater. Process. Technol.* vol. 212 no. 12 pp. 2652–2658.

[13] B. Dilip Jerold and M. Pradeep Kumar 2011 Experimental investigation of turning AISI 1045 steel using cryogenic carbon dioxide as the cutting fluid *J. Manuf. Process.* vol. 13 no. 2 pp. 113–119.

[14] B. Dilip Jerold and M. Pradeep Kumar 2012 Experimental comparison of carbon-dioxide and liquid nitrogen cryogenic coolants in turning of AISI 1045 steel *Cryogenics (Guildf)*. vol. 52 no. 10 pp. 569–574.

[15] C. M. A.-C. Raymond H. Myers and Douglas C. Montgomery 2009 Response Surface Methodology - Process and Product Optimization using Designed Experiments, Third. Canada: *John Wiley & Sons, Inc.*

[16] D. Zhu, X. Zhang, and H. Ding 2013 Tool wear characteristics in machining of nickel-based superalloys Int. *J. Mach. Tools Manuf.* vol. 64 pp. 60–77.

[17] M. Aruna and V. Dhanalakshmi 2012 Optimisation of turning parameters of Inconel 718 alloy using RSM Int. *J. Manuf. Technol. Manag.* vol. 25 no. 1/2/3 pp. 113–134.

[18] S. L. C. Ferreira, R. E. Bruns, H. S. Ferreira, G. D. Matos, J. M. David, G. C. Brandão, E. G. P. da Silva, L. A. Portugal, P. S. dos Reis, A. S. Souza, and W. N. L. dos Santos 2007 Box-Behnken design: An alternative for the optimization of analytical methods Anal. Chim. Acta vol. 597 no. 2 pp. 179–186.

IOP Conf. Series: Materials Science and Engineering **606** (2019) 012016 doi:10.1088/1757-899X/606/1/012016

Structural Analysis of Laterally Aerated Moving Bed (LAMB) Dryer by using Robot Structural Analysis (RSA) Professional 2018.

K Marcellus[1, 2], J Jidon[1, 3], J S Nancy and Y H L Willey[2]

[1] Industrial Dryers and Palm Oil Recovery Research Group, Faculty of Engineering, Universiti Malaysia Sabah, Jalan UMS, 88400 Kota Kinabalu, Sabah, Malaysia.
[2] Mechanical Engineering Programme, Faculty of Engineering, Universiti Malaysia Sabah, Jalan UMS, 88400 Kota Kinabalu, Sabah, Malaysia.
[3]Chemical Engineering Programme, Faculty of Engineering, Universiti Malaysia Sabah, Jalan UMS, 88400 Kota Kinabalu, Sabah, Malaysia.

E-mail: *marcelluskan@gmail.com*

Abstract. The structural design of Laterally Aerated Moving Bed (LAMB) dryer is subjected to buckling effects due to the loads of the paddy bed, self-weight and static pressures by the air blower to overcome the air resistance in the paddy bed, perforated tube and the bed chamber perforated wall. The dryer structural type is a spatial complex structure where the boundary conditions are not very clear, thus the Robot Structural Analysis (RSA) Professional 2018 is used to determine the critical values of loads when the overall structural elements instability occurred to form buckling. The RSA analysed the global buckling of a structure to find the critical coefficient value to be multiply with the case load for the critical buckling load determination. Three different model of dryer structural designs were prepared and analysed by RSA to find the buckling critical coefficient values (α_{cr}), displacements (mm), reactions (N) and moments (Nm). All structural design model was designed with similar type of materials, different sizes and geometrical arrangements. The RSA results within the three structural design models were compared. The structural design model with the best buckling coefficient values with minimum displacement had been selected for the actual LAMB dryer structural construction.

1. Introduction

The Laterally Aerated Moving Bed (LAMB) dryer is a new design for paddy rice drying based on the bioreactor concept which have a vertical perforated tube to provide uniform air flow distribution. The principle of the system is to achieve drying uniformity by aerating the paddy bed laterally. The short air flow distance across the paddy bed produces lower aeration energy contributes to the smaller air flow pressure drops. It has many advantages against the existing industrial paddy rice dryers such as good drying uniformity, low labour intensive, less dust pollution, good heat and mass transfer ability which shorten the drying period *(Nor Hidayah Kamin and Jidon Janaun 2017)*.

The structural analysis of the LAMB dryer structure is vital to ensure it has adequate strength to withstand the loads of paddy bed and pressure inside the dryer vessel. The bed chamber perforated wall and tube are the most crucial areas as there have direct contact with the paddy and pressure from the air blower. The structures are prone to instability due to the loads compression where it starts to change shapes or deflection. Sudden deflection of the structures caused by the compressive load generates the critical stress commonly known as buckling phenomenon *(Helbig D., et al. 2016)*. The structural analysis software such as the RSA is useful to determine the critical values of loads on complicated spatial structures where the boundary conditions are not clear. The structural analysis result obtained by the RSA software is less than 5% difference compared to the result obtained from the analytical way which is acceptable. The critical force, N_{cr} determined by the RSA is always on the safe way *(Chen and Wierzbicki 2000)*.

The selection of materials for LAMB paddy dryer must considers the abrasiveness of the paddy rice husk. The paddy husk contains high silica that causes excessive wear to parts of processing machine *(IRRI n.d.)*. The fabrication of column paddy dryer, galvanised steel was been used for the dryer construction *(Olaniyan and Alabi 2014)*. The fixed bed dryer (Flatbed & Inclined bed dryers) perforated floor and frames were made of mild steel *(Wimberly 1983)*. The plug-flow fluidised-bed paddy dryer was made of stainless steel and mild steel *(Khanali et al. 2012)*. Thus, the materials selected for the LAMB paddy dryer are stainless steel 304 sheet ASTM A240 for the dryer vessel walls, square hollow steel (SHS) and rectangular hollow steel (RHS) ASTM A500 for the dryer structural frames to reduce the excessive wear problems.

The LAMB paddy dryer's capacity of 1000kg was selected for this study. Paddy rice drying analysis by the LAMB dryer was performed to determine the parameters needed for the loads analysis such as the heater power, air flow rate, static pressure and the mass of paddy. These load parameters were used in the RSA to determine the buckling critical coefficient values (α_{cr}), displacements (mm), reactions (N) and moments (Nm).

The RSA is capable to identify any errors on the structural joints due to the nodes instability which is very helpful to detect any design faults of the particular structures. It could analyses the global buckling of a structure to find the critical coefficient values to be multiply with the case load for the critical buckling load determination. In EN1991-1-8 (2005): Eurocode-3: Design of Steel Structures (Part 1-8) standard, the buckling critical coefficient values should be, $\alpha_{cr} \geq 10$ (elastic analysis) and $\alpha_{cr} \geq 15$ (plastic analysis) *(Autodesk Inc. 2012)*.

The LAMB dryer is consisted of two main structures namely the dryer vessel and support chute. In this research, the dryer structural designs were prepared in three different models and analysed by the RSA for comparison. All model was designed with similar type of materials with different sizes and geometrical arrangement to suit the fabrication and installation processes. The vessel structure design models are shown in the Figures 1, 2 and 3 while the support chute is shown in Figure 4. The RSA final results on buckling coefficient values were compared where the design model with the best buckling critical coefficient values was selected for the dryer structures construction. On the support chute structural design, it is confirmed the RSA results obtained in compliance with the Eurocode-3 buckling critical coefficient value, $\alpha_{cr} \geq 10$ (elastic analysis) and $\alpha_{cr} \geq 15$ (plastic analysis).

In conclusion, the Robot Structure Analysis (RSA) professional 2018 package is capable to analyse the buckling effects on LAMB dryer structural analysis.

2. Paddy drying analysis

This analysis is to determine paddy mass and internal pressure needed for the loads analysis. Given, the Inner pipe diameter, D_1 = 0.64m, Paddy Bed diameter, D_2 = 1.17m, Outer wall diameter, D_0 = 1.97m, Dryer height, H = 3.00m, Dryer designed weight capacity, W_p = 1000kg, Bulk density of paddy = 495 kg.m^3, Specification for the perforated sheer / wall: Perforate (Aperture), Diameter of one exit hole = 0.003m, Number of holes (Aperture) = 43.1 holes per inch^2 or 66805 holes per m^2, Area of holes = 0.4722 m^2 holes per m^2 of sheet.

2.1. Paddy mass

The designed capacity of the LAMB dryer is 1000kg, thus the maximum mass of paddy for the load analysis is 1000kg.

2.1.1. Heater power, blower air flow rate and static pressure
The quantity of heat required to remove moisture content from the paddy was determined by using the following equation *(A K Stephen et al 2009)*;

- Heater Power, $Q = M_w \times C_p \times \Delta T$

where, M_w = Mass of water (kg), C_p = Specific heat capacity of water ($kJkgK^{-1}$) and ΔT = Temperature dried grain – Initial temperature of dryer (°C).

2.1.2. Heater power
At $\Delta T = 35°C$, $M_w = 236.62kg$ and specific heat capacity of water of 4.182 $kJkgK^{-1}$. The quantity of heat required to remove moisture content;

- $Q = M_w \times C_p \times \Delta T = (236.62kg) \times (4.182\ kJkgK^{-1}) \times (35) = 3.50 \times 10^4\ kJ$
- The quantity of power = Quantity of heat (Q) / time (s) = 3.241 kJs^{-1} or 3.241 kW
- By giving the heater safety factor of 1.3, the heater power = <u>4.212 kW</u>

2.1.3. Blower air flow rate
Given, the specific heat of air = 1.005 $kJkgK^{-1}$, Power of heat on air = Mass Flow Rate of air x specific heat of air x ΔT, Mass air flow rate = (4.0 kJs^{-1}) / {(1.005$kJkgK^{-1}$ x 35)} = 0.11372 kg. s^{-1}, Density of air at 40°C = 1.127 kg.m^{-3}, Specify volume of air at 40°C = 1 / 1.127 = 0.88731 m^3kg^{-1}

- Air flow rate = Mass flow rate x specific volume at 40°C
  = (0.11372 kgs^{-1}) x (0.88731 m^3kg^{-1}) = 0.101 m^3s^{-1}
- Converting the value of air flow rate to cubic foot per minute (cfm); -
- 1cfm = 4.91747 x 10^{-4} m^3s^{-1}, thus 0.101 m^3 = <u>205.40 cfm</u>

2.1.4. Pressure resistance in the paddy bed
Air flow resistance in the paddy bed is determine using the equation by *(ASABE 2011)*;

$$\frac{\Delta P}{L} = \frac{aQ^2}{\log_e(1+bQ)}$$

where; ΔP = Pressure Drop [Pa], L = Depth of packed bed [m], *a* & *b* = constants for particular grain

- *Q* = air flow, [m^3/s.m^2] or [cfm/ft^2]
- Cross-section of paddy bed = <u>0.753m^2</u>
- Air flow, Q = (air flow rate) / (cross-section area of the bed) = <u>0.14077 ms^{-1}</u>
- Given values of, a = 2.57 x 10^4 [Pa. s^2/m] and b = 13.2 [m^2. s/m^3] *(for rough rice)*.
- Pressure drop, $\Delta P_B = [(2.57 \times 10^4)\ (0.14077)^2(3)]\ /\ \text{Log}_e\ \{1 + (13.2)\ (0.14077)\}$ = 3,349.765 Pa

The air resistance in the paddy bed = Pressure drop in the paddy bed = <u>3.35kPa</u>

2.1.5. Air flow resistance through the perforated tube walls

Given, the porosity of the perforated walls, *f* = 46.47%. The open area ratio (porosity, *f*) to be used on the following equations to find the *k* -factor to determine the pressure drop *(Idel'chik 1994)*.

$$k = [0.707(1-f)^{0.375} + 1 - f]^2 \frac{1}{f^2} \quad and \quad \Delta P = k\left(\frac{\rho v^2}{2}\right)$$

- f = porosity, ΔP = pressure loss [Pa], v = velocity ($m.s^{-1}$), ρ = density of air at 40°C = 1.125 $kg.m^{-3}$
- $k = [0.707(1\text{-}0.46467)^{0.375} + 1 - 0.46467]^2\,[1/(0.46467)^2]$ = <u>5.5495</u>
- v = velocity ($m.s^{-1}$) = Air flow rate / Open area segment of inner pipe = <u>0.037 $m.s^{-1}$</u>
- Thus, pressure drop, ΔP_1 = 5.5595 (1.127 $kg.m^{-3}$)(0.037 ms^{-1})^2 / 2 = 0.116 $kg.m^{-1}s^{-2}$= <u>0.116 Pa</u>

2.1.6. Air flow resistance through the perforated bed wall
Given, k = 5.5495, ρ = density of air at 40°C = 1.125 $kg.m^{-3}$

- v = velocity ($m.s^{-1}$) = Air flow rate / Open area segment of paddy bed = <u>0.0204 $m.s^{-1}$</u>
- pressure drop, ΔP_2 = 5.5595 (1.127 $kg.m^{-3}$)(0.0204 ms^{-1})^2 / 2 = 0.064 $kg.m^{-1}s^{-2}$ = <u>0.064 Pa</u>

2.1.7. Static pressure
The static pressure of the blower = pressure resistance in the paddy bed + pressure drop on the perforated tube + pressure drop on the bed perforated wall

- Total Pressure Drop, $P_s = \Delta P_B + \Delta P_1 + \Delta P_2$ = 3.35 x10^3 + 0.116 + 0.064 = 3,350.18 Pa
- Given the safety factor of 1.3, the blower static pressure = <u>4.36 x 10^3Pa</u>

2.2. Loads analysis
The loads on the dryer vessel frame were contributed by the vertical force due to the mass of paddy and the horizontal forces caused by the internal pressure of the vessel.

2.2.1. Vertical loads
The forces are determined by Newton's second law (F = mg), where m = mass (kg) and g = gravity acceleration (9.8 m/s^2). The vertical forces were caused by the load of the paddy mass from *(1.1.)*.

- Vertical force, F_z = 1000 kg x 9.8 m/s^2 = 9.8 $kg.m/s^2$ = 9.8 kN ≈ 10kN

2.2.2. Horizontal loads
Horizontal forces are determined by F = PA, where F = horizontal force (N), P = pressure (Pa or N/m^2) and A = surface area (m^2).

2.2.2.1. Horizontal load on the inner perforated tube
Perforated tube wall surface area = circular perimeter x height = 1.005 m^2

- $F_{x1} = F_{y1}$ = (1.005 m^2) x (4.36 x 10^3) = 4381.8N or 4.40 kN

2.2.2.2. Horizontal load on the paddy bed wall
Bed perforated wall surface area = circular perimeter x height = 1.838 m^2

- $F_{x2} = F_{y2}$ = (1.838 m^2) x (4.36 x 10^3) = 8013.68N or 8.01kN

2.2.2.3. Horizontal load on the outer wall
The pressure at the outer wall is merely the residual pressure through the bed perforated wall.

- The pressure on the outer wall = (4.36 x 10^3) – (3.35 x 10^3) = 1.01 Pa
- The outer wall area = 9.283m^2
- $F_{x0} = F_{y0}$ = (9.283m^2) x (1.01Pa) = 9.376N ≈ 0.01kN

3. Structural analysis

Three structural models were prepared for the LAMB dryer vessel and analysed by the RSA. The RSA results were then compared to select the best structural strength for the LAMB dryer structures construction. Given; the vertical load, F_z = 10kN and the horizontal loads; $F_{x0} = F_{y0}$ = 0.01kN, F_{x1} F_{y1} = 4.40kN and $F_{x2} = F_{y2}$ = 8.01kN *(In the RSA, the horizontal loads shall be in x and y axis, the vertical loads shall be in z axis.).*

3.1. Model Structure No. 1

The selected materials for model no.1 are RHS 50x25x2.5 and SHS 50x50x3 as shown in Figure 1.

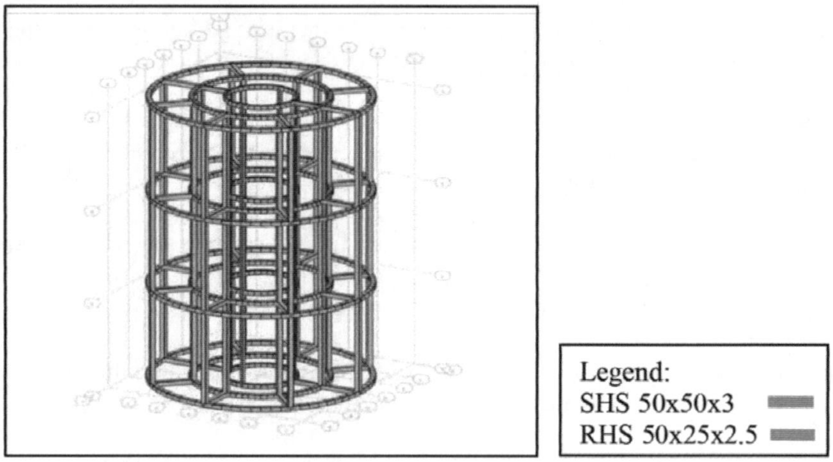

Figure 1. Model Structure No. 1

3.1.1. RSA results for Model No. 1

Table 1. Buckling Analysis Results - Values (Model No. 1)

Case/Mode	Critical coefficient	Precision
2/1	6.12020e+00	3.05759e-04
2/2	6.79447e+00	3.37184e-03
2/3	8.51962e+00	1.77585e-03
2/4	8.91491e+00	3.84918e-03

3.2. Model Structure No. 2

The selected materials for model no.2 are RHS 50x25x2.5, SHS 50x50x3 and SHS 25x25x2.5 as shown in Figure 2.

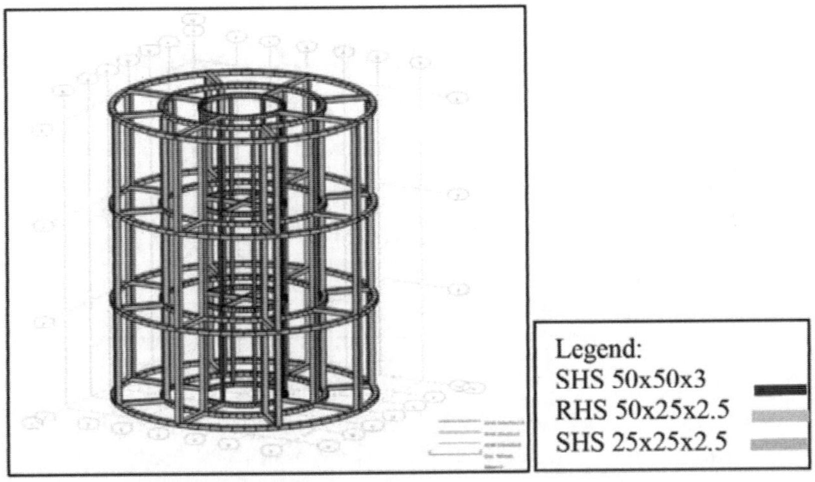

Figure 2: Model Structure No. 2

3.2.1. RSA results for Model Structure No. 2

Table 2. Buckling Analysis results - Values (Model No. 2)

Case/Mode	Critical coefficient	Precision
2/1	3.94137e+00	1.12316e-03
2/2	3.94298e+00	1.12086e-03
2/3	4.12656e+00	1.16044e-03
2/4	4.13121e+00	3.29510e-04

3.3. Model Structure No. 3

The selected materials for model no.3 are SHS 50x50x3 and SHS 40x40x2.5 as shown in Figure 3.

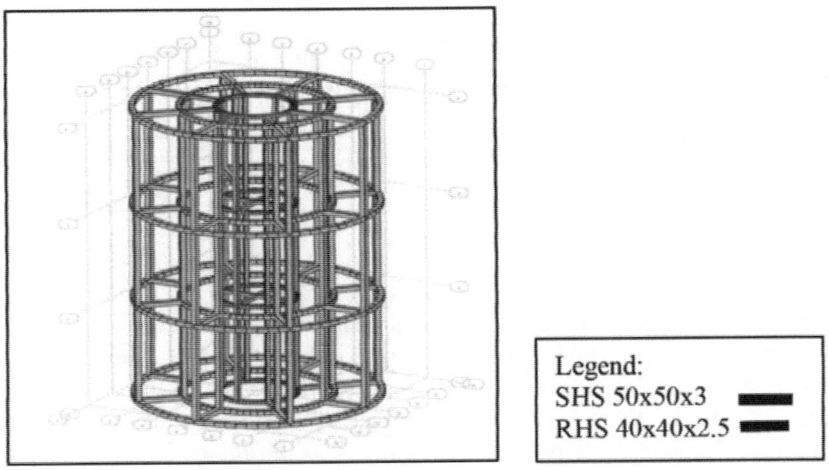

Figure 3. Model Structure No. 3

3.3.1. RSA results for Model Structure No. 3

Table 3. Buckling Analysis Results - Values (Model No. 3)

Case/Mode	Critical coefficient	Precision
2/1	1.67156e+01	2.40375e-03
2/2	1.67158e+01	3.66079e-03
2/3	1.88446e+01	1.66991e-03
2/4	1.88446e+01	2.43442e-03

3.4. RSA for support chute

The support chute is made of S275 steel, CRS 152x102x23.07, SHS 50x50x3, UB 152x89x16 and S275 steel plate of 8mm thickness. The vertical load, F_z = 25kN (total loads of the mass of paddy and maximum weight of the dryer vessel).

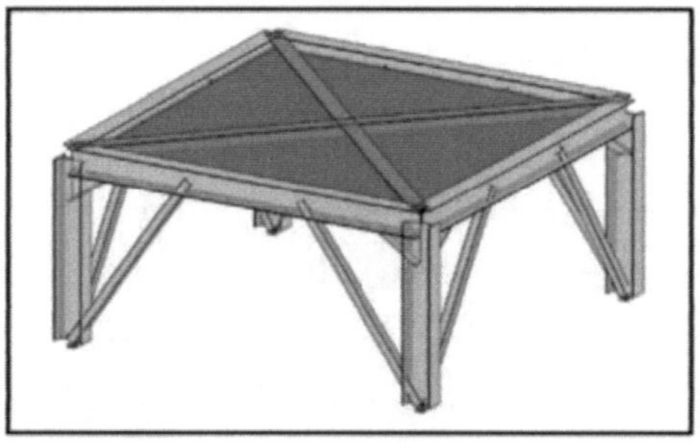

Figure 4. Support Chute Structure

3.4.1. RSA results on the support chute

Table 4. Buckling analysis results values

Case/Mode	Critical coefficient	Precision
2/1	2.08568e+01	9.23010e-07
2/2	1.13765e+01	2.20282e-03
2/3	1.15870e+01	1.13529e-03
2/4	1.16739e+01	1.85424e-03

4. Discussion

Three structural design models for the dryer vessel were analysed by the RSA to find the buckling analysis critical coefficient values. The RSA buckling results obtained on the critical coefficient values for design structural Model No.1 and No.2 as tabulated in Table 1 and 2 respectively are found less than 10.

The structural design Model No.3 RSA buckling results on critical coefficient values as shown in Table 3 are found to be in compliance with the EN1991-1-8 (2005): Eurocode-3: Design of Steel Structures (Part 1-8) standard, the buckling critical coefficient values should be, $\alpha_{cr} \geq 10$ (elastic analysis) and $\alpha_{cr} \geq 15$ (plastic analysis).

The buckling analysis results on critical coefficient values for the Support Chute as shown in Table 4 are also found to be in compliance with the EN1991-1-8 (2005): Eurocode-3: Design of Steel

Structures (Part 1-8) standard, the buckling critical coefficient values should be, $\alpha_{cr} \geq 10$ (elastic analysis) and $\alpha_{cr} \geq 15$ (plastic analysis).

5. Conclusion

The Robot Structural Analysis (RSA) Professional 2018 package is capable to perform structural analysis for the Laterally Aerated Moving Bed (LAMB) dryer.

The best structural model design for the LAMB dryer vessel is Model Structure No.3 as shown in Figure 3 and Table 3 to be used for the actual construction of the dryer vessel structure.

Further studies and experiments will be carried out upon completion of the LAMB dryer construction for processes validation, cost and performance optimisation, serviceability and sustainability.

Acknowledgement

The authors would like to thank Villaco Sdn Bhd for the commercialisation of LAMB dryer and Universiti Malaysia Sabah for providing the financial support (SGI0058-2018).

References

[1] A K Stephen et al. 2009. 2 *Improvement on the Design of a Cabinet Grain*. ed. American J. of Engineering and applied sciences.

[2] ASABE. 2011. 1996 *ASAE D272.3 DEC01 - Resistance to Airflow of Grains , Seeds , Other Agricultural Products , and Perforated Metal Sheets.*

[3] Autodesk Inc. 2012. Autodesk, Inc. *Autodesk Robot Structural Analysis - Verification Manual According to AFNOR*. 5.2. USA: Autodesk Inc.

[4] Chen, Weigang, and Tomasz Wierzbicki. 2000. "Torsional Collapse of Thin-Walled Prismatic Columns." *Thin-Walled Structures* **36**(3): 181–96.

[5] Helbig D., et al. 2016. "Study about Buckling Phenomenon in Perforated Thin Steel Plates Employing Computational Modeling and Constructal Design Method." *Latin American Journal of Solids and Structures* 13(10): 1912–36.Idel'chik, I E. 1994. "Handbook of Hydraulic Resistance (3rd Edition)." *Washington*: 517.

[6] IRRI. Agricultural Engineering International *Rice Milling*. http://www.knowledgebank.irri.org /ericeproduction/PDF_&_Docs/Teaching_Manual_Rice_Milling.pdf.

[7] Khanali, M et al. 2012. "Mathematical Modeling of Fluidized Bed Drying of Rough Rice (Oryza Sativa L.) Grain." *Journal of Agricultural Technology* 8(3): 795–810.

[8] Nor Hidayah Kamin and Jidon Janaun. 2017. "Drying of Paddy in a Laterally Aerated Moving Bed Dryer at Ambient Temperature." *Chemical Engineering Transactions* **56**: 877–82.

[9] Olaniyan, Adesoji Matthew, and Abiola Adejoke Alabi. 2014. "Conceptual Design of Column Dryer for Paddy Rice Including Fabrication and Testing of Prototype." *IJBAC* 02(03): 11–20.

[10] Wimberly, James E. 1983. IRRI *Technical Handbook for Paddy Rice Postharvest Industry in Developing Countries*.

Institute of Physics
Dirac House, Temple Back
Bristol BS1 6BE UK

ISSN: 1757-8981
ISBN 978-1-5108-9483-9

CURRAN ASSOCIATES INC.
proceedings.com